Full Fathom 5000

Full Fathom 5000

The Expedition of HMS Challenger and the Strange Animals It Found in the Deep Sea

GRAHAM BELL

OXFORD
UNIVERSITY PRESS

OXFORD
UNIVERSITY PRESS

Oxford University Press is a department of the University of Oxford. It furthers
the University's objective of excellence in research, scholarship, and education
by publishing worldwide. Oxford is a registered trade mark of Oxford University
Press in the UK and certain other countries.

Published in the United States of America by Oxford University Press
198 Madison Avenue, New York, NY 10016, United States of America.

Library of Congress Control Number: 2021044997

ISBN 978-0-19-754157-9

DOI: 10.1093/oso/9780197541579.001.0001

1 3 5 7 9 8 6 4 2

Printed by Sheridan Books, Inc., United States of America

Full fathom five thy father lies;
Of his bones are coral made;
Those are pearls that were his eyes;
Nothing of him that doth fade
But doth suffer a sea-change
Into something rich and strange.

Shakespeare, The Tempest, Act 1, Scene 2

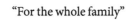
"For the whole family"

Contents

Acknowledgments

I am extremely grateful for the help given by Lauren Williams and the Rare Books Collection of McGill University in supplying high-quality images from the Reports, which are the basis for the figures in this book. I benefited greatly from visits to the Caird Library of the National Maritime Museum, Greenwich, and the Foyle Reading Room of the Royal Geographical Society in South Kensington.

A Note on Names and Units

I have avoided common names for animals wherever possible. The reason is that they are rarely used in common speech. There are exceptions, such as crab and starfish, but who would ever refer in conversation to a comma shrimp or a sea lily? I have instead used the zoological names (cumacean and crinoid, in this case), even though they may be new terms to most readers, on the grounds that any terms will be new and these are exact. It is a little like getting to know the characters in a novel; the names themselves are not important, but you need to know them to understand the plot. I have explained the standard system for naming individual species in the text (Station 106, 25 August 1873). The names of animals sometimes change, however, and the names given in the contemporary Reports may not correspond with those currently accepted; I have taken the World Register of Marine Species (WORMS) as being authoritative.

The most important quantity in the book is depth, which I have given in fathoms, as in the original documents of the voyage. A fathom is six feet, or about two meters. Multiplying the depth in fathoms by two will give you the depth in meters, to an acceptable approximation. Most distances are given in miles, about 1.6 kilometers, or nautical miles, somewhat more; the unit of velocity is the knot, which is one nautical mile per hour or about 1.85 kilometers per hour. Other dimensions, of animals for example, are given in whatever unit seems most appropriate; it may be useful to recollect that one inch is about 25 millimeters and one foot about 30 centimeters. False precision is the enemy of understanding.

Introduction

If you're the sort of person who reads introductions (which you are, obviously), I thought you might like to know what this book is about and how it came to be written. I teach biology at McGill University in Montreal, including a rather old-fashioned course on zoology in which I describe all the main groups of animals. Most of these live in the sea. There are plenty of animals on land, of course, but almost all of them are insects or vertebrates, plus a few snails and spiders. Because there are many more different kinds of animal in the sea than on land, I found myself preparing lectures by reading a lot about marine biology, despite not being a proper marine biologist. It was not long before I began to come across references to the voyage of HMS *Challenger*, back in the 1870s, when many of these animals were collected for the first time. It was obviously a famous affair. The newspapers of the day printed progress reports, and the officers were greeted by royalty, or at least the nearest local equivalent, at many of the ports they visited. All this public attention was because the voyage had a unique objective: it was a scientific expedition to find out what (if anything) lived at the bottom of the deep sea. Nobody knew for sure. Biologists had paddled at the edge of the sea since Aristotle, but anything living deeper than the handle of a net was for all practical purposes out of reach. A few animals were brought up from time to time by fishing gear or ships' anchors, but otherwise the only people to visit the bottom of the sea were dead sailors. The first sustained attempts to explore this unknown world were not made until the Industrial Revolution was well under way. When the *Challenger* expedition sailed in 1873, it was to make the first systematic investigation of what lay beneath the surface of the world's oceans. Nobody knew what it would find. Anecdotes aside, nobody was sure what covered the sea floor, or what lived there, or even how deep it was. It was the Victorian equivalent of a voyage to the surface of the moon.

The voyage was not particularly eventful, in fact. There were no battles at sea, no shipwrecks, no mutinies to be quelled or pirates to be fought off. It was fairly comfortable, at least by nineteenth-century standards, and the crew were never reduced to eating rats or boiled boot-leather. The scientists

Full Fathom 5000. Graham Bell, Oxford University Press. © Graham Bell 2022.
DOI: 10.1093/oso/9780197541579.003.0001

on board were very distinguished, but their names and reputations have long since faded into the Victorian mists. There were no women on board at all, no affairs and no scandal. So what is there to write about? Not surprisingly, most of the narrators have chosen to describe at length the time spent on shore, especially the visits to exotic and unfamiliar places, and tend to gloss over the time spent merely sailing from one port to another. Their accounts make very interesting reading as Victorian travelogues, but it seemed to me that something was missing—such as the main point of the expedition, the animals that it found in the deep sea.

That's why I began to make notes about the animals that had been discovered during the expedition, and then to trace the voyage on a very large map, and then to scrutinize the species lists for each station, and by then it was too late; at some point it became easier to write the book than not. But why are the animals so important? Well, I suppose that for convenience you might recognize three kinds of animal: there are those you can see, on land and in shallow water; there are those you can't see, because they are too small; and then there are those you can't see because they are hidden from sight in deep water. The first kind is familiar to us all; the second kind was discovered by the early microscopists; and the third kind was discovered by the *Challenger* expedition. Most of the species captured from the deep sea during the expedition had never been seen before, either by scientists or by anyone else. The expedition did not merely lengthen the catalog of living animals, but, much more than that, added a whole new volume to accommodate the hidden fauna of half the world.

There are people on the stage too, of course, especially the scientists on board. Their leader was the portly and somewhat pompous Charles Wyville Thomson, accompanied by Henry Moseley, who always seems a little raffish; the saturnine chemist John Young Buchanan; and the earnest student Rudolf von Willemoes-Söhm, who was drafted more or less by accident. John Murray belongs in a separate sentence as the ablest of them, the deepest thinker, the hardest worker, and the only one of them to turn a profit from the voyage. Then there were the officers: the two captains, George Nares and Frank Thomson; John Maclear and Tom Tizard, who did most of the navigating; the aristocratic lieutenant Lord George Campbell, who left the raciest account of the voyage; his junior Herbert Swire, who left the grumpiest account but unfortunately bowdlerized it at the last minute; and all the others needed to work a naval ship. There were also 200 or so anonymous seamen in the background, which is where they stay, as usual, with the peculiar exception of Assistant Steward Henry Matkin, whose letters home have survived and give us a rare glimpse of life on the lower deck.

The voyage itself is the thread on which the animals are strung as we pass from station to station across the oceans for nearly 80,000 miles. I have given short shrift to the visits on shore, which can be read about in other books, but to spare the reader I have also omitted a lot of technical stuff about currents and sediments that seemed to me less than gripping. What remains is the animals themselves, including the ugliest fish in the world, flesh-eating clams, dwarf males, sea devils, and an octopus that wears lipstick. I hope that you will be as fascinated by these strange creatures of the deep sea as I have been, as we follow HMS *Challenger* on her long and complicated voyage around the world.

PART 1
BEFORE

1

The Deep Sea

Mudskippers are small, popeyed fish that hop around on beaches in the tropics, foraging for anything that crawls just above the high-tide mark. They are on the land but not of the land, so to speak, tourists rather than residents, who can venture no further than the mangrove roots. We are mudskippers in reverse, familiar enough with the bright rim of the sea and occasionally venturing a little further with mask and snorkel, but unable to leave the shore without the sort of technology that mudskippers lack. Even with our best technology, however, we are no more than tourists on an extended visit. The mudskipper sees the beach and the fringe of the forest, but cannot imagine mountain, desert, and river. We see the shore and the reef without realizing that a mile or two beyond lies a very different place, in which none of the rules by which we are accustomed to live our lives hold any more.

If you could walk directly into the sea, through the surf and away from the land, you would be likely to descend a gentle slope for two or three days. The water over your head would gradually become deeper, until the surface was about 200 meters above, say two average city blocks away. About this time the gradient would begin to increase, and soon you are walking quite steeply downhill, with the surface rapidly receding. The sediment beneath your feet will thicken as you approach the edge of the continent, where the soil and debris washed from the land eventually accumulates. At the end of a long day's walk the gradient eases and you stride out onto a prairie that stretches away into the far distance, flat and featureless, or sometimes with rolling hills or even studded with sudden abrupt mountains. The surface is now quite far away, not mere city blocks but the whole downtown core of a large city distant from where you stand. You have arrived at the edge of the abyssal plain. The details of your walk will vary according to where you start out, and might be quite different if you begin in eastern Canada and walk into the Atlantic, say, or in western Canada and walk into the North Pacific. In either case, though, you will eventually reach the abyssal plain, where the last of the land has been left behind and nothing is familiar.

Full Fathom 5000. Graham Bell, Oxford University Press. © Graham Bell 2022.
DOI: 10.1093/oso/9780197541579.003.0002

To begin with, it is dark. If water were as transparent as air, every detail of the bed of the sea would be as clear as the hills and valleys seen from an airplane. But water seems transparent only in small quantities, in a glass or a bath. You can see pebbles and fish 100 feet down in water as pure as Georgian Bay or the Sulu Sea, but beyond this light fades even in the purest seas. Beyond the sunlit zone, just a few hundred feet down, there is no light. It is always nighttime, but darker than night; the darkness is absolute and perpetual, without sun or moon or stars. Without sunlight, there are no forests or meadows, but in their place a vast bare plain, relieved only by the occasional thicket of sponges or gorgonians. Without forests and meadows, there are no plants to eat, but instead only the rain of dead bodies from the lighted zone, a perpetual thin snow of bacteria, fecal pellets, and unfamiliar organisms such as coccolithophores and copepods, chaetognaths and ctenophores, with the occasional plunging body of a dead whale or shark.

It is also very cold; it is always winter in the depths of the sea. The water is everywhere cold, as cold as snow melt, although it never freezes. It is very calm, far from surf, wave, and tide, stirred only by an almost imperceptible current. There are ridges and ravines in some places, but most of the seabed is flat or gently sloping and floored with an impalpable ooze long smoothed by gravity. The pressure of the long column of water is crushing. Each 10 meters of water adds one atmosphere of pressure, and in the deep sea the pressure is 100 times, or more, what we are accustomed to on land. A can of beans would implode; a strong steel safe, carefully sealed, would be bent and buckled; wood is crushed, so that knots stand proud of the surface.

Deep water is not just different but different in kind from the edge of the sea, just as deep space and deep time cannot be understood by simply extrapolating from mundane scales. At first glance, the abyssal plain might seem to be a hostile and extreme environment, where darkness, cold, dearth, and pressure reach their limits in the ocean. And yet it is by no means an exceptional environment. Quite the contrary: abyssal plains cover a little more than half the surface of the Earth. Far from being the exception, they are the rule: they form by far the largest habitat by area, and the waters immediately above them form by far the largest habitat by volume. Despite their vast extent, we know very little about them. Nobody can live down there, of course, and even a brief visit requires specially built vessels that only a rich country can afford. Remotely operated vehicles and cameras have been used in recent years to peer into the deep sea, but even today only a few percent has yet been explored. Two hundred years ago we knew nothing at all beyond a few anecdotes. In particular, we knew nothing of the animals that live there, not even whether there were any. Now we know that animals do live in the deep sea, but most are of unfamiliar kinds, or strangely altered versions of familiar kinds. None of them had ever been seen by human eyes, and they lived as they had lived for long ages, unknown, untaken, and secure from our prying gaze—until 1872.

2

Edward Forbes

It all goes back to Forbes, eventually, or at least I think it does. Edward Forbes. His name is not much remembered nowadays, and his reputation has been thoroughly plowed under by modern commentators. You might say, quite fairly, that he had only two big ideas, and one of them was wrong. In his defense, having even one big idea that is scientifically credible is far from commonplace—his detractors have not managed one yet—and his errors were more fruitful than a hundred narrow truths. Here is how it all began.

On the evening of 23 February 1844 Edward Forbes read a paper at the meeting of the Philosophical Society of Edinburgh, now the Royal Society of Edinburgh. It was about dredging. This is not a very promising subject; it sounds somewhere between "dreadful" and "grudging." It is not a very romantic operation, either, at least on the face of it. A dredge is a net attached to a rigid frame, towed along the seabed for a while before being brought onboard for its contents to be examined. Pause for a while to consider: you are in a small boat on a rough sea, pulling on a couple of hundred feet of slimy rope to bring in a net full of mud. It is not an enchanting prospect. Well, put it another way. Most of the Earth's surface is under sea, and its inhabitants are unknown and inaccessible to the creatures who live on the land. Forbes's feeble scratchings were the first systematic attempt to find out what creatures lived far, far below the surface of the sea.

Forbes was born at Douglas on the Isle of Man in the Waterloo year of 1815, and he spent the first 15 years of his life there. He seldom attended school and instead passed his time poking around in the local bogs and beaches, a naturalist in the making. By the age of 16 he was familiar with the fauna of the island and had even visited an offshore bank where the fishermen used a dredge to bring up scallops together, inevitably, with a variety of other animals. At this point he left home for Edinburgh, where he was intended to take a medical degree at the university. He could scarcely have chosen better. In the late eighteenth and early nineteenth centuries the Scottish universities, Edinburgh and Glasgow in particular, led a new revival of learning, at a time when Oxford and Cambridge were moribund. They produced a crowd

Full Fathom 5000. Graham Bell, Oxford University Press. © Graham Bell 2022.
DOI: 10.1093/oso/9780197541579.003.0003

of illustrious names: David Hume in philosophy, Adam Smith in economics, James Boswell in literature, Joseph Black in chemistry, James Hutton in geology, Colin Maclaurin in mathematics, and a host of others. Forbes had come, not to a provincial college, but to the New Athens.

As a young man, Forbes was somewhat slight in build; he was clean-shaven and wore his hair long, almost to his shoulders. His face was lean and saturnine with a slightly melancholic cast and rather protuberant eyes. Nevertheless, he had a cheerful disposition and made many fast friends, including John Goodsir, who shared his interest in marine life and sailed with him many times when they were students together. His letters and manuscripts are sprinkled with humorous sketches and comic songs; he seems to have been a congenial companion and a "wine-bibber," as he refers to himself in a letter to his brother. He was also clearly an active and vigorous man, well able to handle a small boat in a rough sea. Yet there was some physical frailty that led to recurrent serious illness. He was kept at home until he was 11 years old by his parents, because of some childhood illness or indisposition, and later had two alarming episodes, a high fever that kept him in bed for over a week at sea far from medical help, and later an attack of scarlet fever, a serious matter in the days before antibiotics, while fossil-hunting in Dorset.

He was enrolled in the medical school, which at that time could claim to be one of the foremost scientific institutions in Europe. It didn't take. He seems to have dutifully attended the lectures of the medical curriculum, but his heart was not in it, and he eventually abandoned his studies without taking a degree. What he did not abandon was his enthusiasm for natural history. One of his professors was Robert Jameson, the Regius Professor of natural history for almost the entire first half of the nineteenth century. Jameson's most famous pupil was Charles Darwin, who found his lectures tedious, but his most dedicated pupil was Edward Forbes, who didn't, and who was, in the end, destined to be his successor. Jameson was a museum man, in the days when the principal goal of the biologist was to collect and catalog as many specimens as possible, in order to fill out the plan of the world and thereby to discover the mind of God. Forbes was himself a collector from his earliest youth and must have found this congenial; at all events, he turned away from his medical studies to concentrate more and more on natural history.

In May 1832 Forbes hired a boat and went dredging in the Firth of Forth, before attending Jameson's lectures. Two years later, in the summer of 1834, he again went dredging, this time on a bank off the coast of the Isle of

Man. He was working in relatively shallow water—20 fathoms or so—perhaps because of the limitations of his gear, but nevertheless far beyond the range of any shore-haunting naturalist. He was not the first to dredge offshore: Otto Muller had dredged in deeper water in the Norwegian fjords as early as the 1770s, and indeed Forbes used a dredge based on Muller's design. Nevertheless, Forbes's tentative trials with primitive gear mark the beginning of an organized and concerted effort to find out what lived on the sea floor. He made detailed notes of the animals he found, mostly limpets, snails, and brittle stars, but also a crinoid that surprised and disappointed him by promptly disarticulating itself. (We shall meet crinoids later; they are not animals that a shore-borne naturalist would ever come across.) He published his results in a respectable scientific journal, the *Magazine of Natural History*, with detailed descriptions of the animals he found and the conditions in which they lived. He was, in short, beginning to carve out a career for himself and, in doing so, to point biology in a new direction.

It is not obvious now how such a mundane activity as towing a net along the bottom of the sea should be such a novel idea. At the time, however, we knew almost nothing about the animals that lived on the seabed far from land. The mirror of the sea concealed them from us. Sailors and fishermen knew something of what swam in the upper waters of the ocean, but their knowledge of the depths was confined to what was brought up on the wax of the sounding lead. It was as though, reversing the roles, we had evolved as marine creatures, able to explore the intertidal zone, but knowing nothing of what lay beyond except for a few blurred glimpses of beach or cliff. Forbes broke the mirror of the sea and began the first systematic exploration of an unknown world.

Forbes relinquished his medical studies at Edinburgh (by the simple expedient of failing his exams) and after 1834 turned his attention exclusively to natural history. At the same time, he began to mature from a rather reclusive young man into an outgoing and influential member of the scientific community. In August 1836 he traveled to Bristol (not an easy journey from Edinburgh, then as now) to attend the annual meeting of the British Association for the Advancement of Science, where he found himself in the company of most of the leading British scientists of the day, together with many of their younger colleagues. He found their company very congenial—so much so, indeed, that he attended every subsequent meeting (except when he was out of the country) for the rest of his life. Every year, he presented a report of his researches, soon became secretary of Division D (Zoology and

Botany), and, in short, became widely known and respected among his peers. Forbes was a coming man.

The trouble was, Forbes had no job; nothing that paid a salary. He was, in fact, supported by his father, a successful banker in Douglas, and otherwise had no income of his own. He went on a couple of summer cruises, paid for by grants from the British Association, but he must have been relieved when his application to join the survey ship HMS *Beacon* as naturalist was accepted. From early in 1841 to the summer of 1842 he cruised in the Aegean and made use of the opportunity to develop his ideas about the distribution of animals in the sea, with all the assistance provided by a naval vessel. He caught plenty of animals in the dredge, and noticed that each interval of depth was inhabited by a characteristic set of species. He concluded that animal communities occupy distinct zones in the sea, comparable with the zonation that can readily be seen between the tidemarks. He also noticed that animals as a whole became fewer and less diverse at greater depths. Not content with a superficial impression, he plotted animal abundance against depth and then extrapolated the plot to zero abundance, marking the depth below which, he predicted, no animals would be found. This was a very early and very clever use of quantitative analysis in ecology, and led him to the conclusion that the limit of animal life in the sea lay at about 300 fathoms. This was Forbes's "azoic hypothesis," which had a great influence on the planning of future dredging expeditions. It has been ridiculed by modern commentators, who point out that there were previous reports of animals from much greater depths: John Ross, for example, had recovered starfish and tubeworms from 1,000 fathoms in the Arctic as early as 1818. It seems to me that this criticism misses the point. These previous reports were anecdotes, not the result of systematic surveys. They were poorly documented; estimates of depth were unreliable; and a few stragglers would not substantially alter Forbes's conclusion anyway. There were comparable anecdotes of sea-serpents, after all, some from witnesses of unimpeachable authority on board naval ships. Forbes was wrong, as a matter of fact—the Aegean has an exceptionally scanty fauna— but his argument was sound and his hypothesis eminently testable.

It was a brilliantly successful cruise that furnished him with the two leading ideas for which he is remembered, but it ended with catastrophic news from home. His father had flourished, as bankers do, but was now ruined, as bankers sometimes are. His remittances were at an end, and he was forced to abandon plans for a further cruise to the Red Sea and instead return to London and find a job.

He began as curator in charge of the museum of the Geological Society of London in Craig's Court, a very curious blind alley a few yards from Trafalgar Square. Most fortunately, the professorship of botany at King's College London also fell vacant at this time, and (with a little help from his friends) Forbes secured this appointment too. The problem was that these positions, although they were his entrée into professional science, and certainly kept him very busy, did not pay very much; only enough together to keep him in stirabout and water-porridge, as he put it. To put butter in his porridge, he moved to the Geological Survey as the head of paleontology in 1844. He was now self-sufficient (especially because he kept his position at King's), and began to clamber further up the academic ladder, soon being elected a Fellow of the Royal Society and giving lectures at the Royal Institution and later at the newly created School of Mines. He organized moving the collections of the Geological Society, which had outgrown their original site, to the new Museum of Practical Geology in Jermyn Street, which was opened by Prince Albert in the Exhibition year of 1851. (In 1935 they were moved again to the Geological Museum, now part of the Natural History Museum, in South Kensington.) His dredging days were over, though, it seemed, and he applied himself mainly to other subjects. He did go on one last summer cruise in the Atlantic in 1850, but after this he turned more toward geology and paleontology, with great thoroughness, but, I think, rather less enthusiasm and originality. His great opportunity to go back to his early love came in the spring of 1854, when Robert Jameson died in Edinburgh and his Regius Chair became vacant. Forbes was hustled into it by May 1854 (I think Huxley was the prime mover) and was free to return to dredging, where he had begun his career in science.

Forbes had now reached his zenith. He had secured one of the most prestigious positions in his field in the country; he was on cordial terms with Charles Lyell, Henry Huxley, and Charles Darwin; he had a growing family— he had married in 1848—and a fresh start in research. He was at the forefront of modern thinking about biology: a few years previously he had given a lecture at the Royal Institution with the potentially controversial title "Have new Species of Organized Beings appeared since the Creation of Man?" His answer was "no," but he was very close to showing that a steady stream of new species had come into being during geological time. In four years' time, Darwin and Wallace would publish the first account of evolution by natural selection, with *On the Origin of Species* appearing in the following year. There is no doubt that Forbes would have been deeply engaged, one hopes on the side of the apes rather than the angels. But by then he was dead. He suffered one last bout of severe illness in November that year and died in a week. He seems to have died from acute kidney failure, which can be a complication of the scarlet fever he had contracted four years previously, but that is just a guess. Had he lived, the history of biology might have been different. After his death, his influence continued to be felt through two committees: the Dredging Committee, which Forbes founded, and its successor, the Circumnavigation Committee, which launched the *Challenger* after Forbes had become little more than a memory.

3

Two Committees

Committees have a bad name, and no doubt most of them deserve it. A committee can be a good excuse for dozing away an afternoon with little preparation and no real responsibility. But there is another sort of committee, which focuses the combined ambitions of a set of people, perhaps not otherwise much in sympathy, to achieve something that none would be capable of performing individually. An effective committee must have three attributes: its members must agree, at least with respect to a particular objective; it must be able to define its objective in practical terms; and it must have access to the money, power, or influence sufficient to achieve its objective. Not many committees are that effective. The *Challenger* voyage happened because two were: the Dredging Committee of the British Association for the Advancement of Science, and the Circumnavigation Committee of the Royal Society.

The Dredging Committee

By the late 1830s Forbes had the influence to push for institutional support of his dredging program. I doubt he met with much resistance, because it was an attractive idea that would have a broad appeal to naturalists. The terrestrial and freshwater fauna and flora of Britain were becoming quite well cataloged by the mid-nineteenth century, and dredging in the sea promised a harvest of many little-known marine organisms, opening up a fresh field for exploration and enquiry. But Forbes went deeper than this. Discovering and naming new species was the favorite occupation of naturalists, but as biology advanced, it became clear that it was not a sufficient end in itself. Forbes addressed more fundamental themes: the geographical distribution of organisms in the sea, and the succession of ecological communities from the shore between the tides to water as deep as his dredges could reach. It was these broader scientific questions that gave Forbes the weight to persuade his colleagues that a formal and (most importantly) funded dredging program

Full Fathom 5000. Graham Bell, Oxford University Press. © Graham Bell 2022.
DOI: 10.1093/oso/9780197541579.003.0004

should be set up by the Association. His colleagues were duly persuaded, and at the 1839 meeting, in Birmingham, a Dredging Committee was constituted, with Forbes, of course, as a member, together with his dredging companion John Goodsir. These youngsters were not trusted to head the committee, however, which was instead chaired by J. E. Gray, a more senior figure from the British Museum. Forbes had nonetheless gained his point: dredging was now publicly recognized as a respectable scientific activity by a prominent scientific organization. More than that, it was an activity worth funding: in its first year, the Dredging Committee received a grant of £60 to support its activities. Enough to be going on with.

The Dredging Committee met annually for several years, and although Forbes could not attend in 1841 and 1842—he was with *Beacon* in the Aegean—he took over the chairmanship of the committee from Gray in 1843. The Dredging Committee itself acted mainly as a general administration that set up local committees to organize dredging in different parts of the British Isles. These local committees were provided with standard blank forms to record the animals they found, with details of the type of deposit, distance from the shore, and so forth, and were funded by the central body. This was admirably systematic, but in practice it did not work very well. One sign that the scheme was not working as intended was the failure to spend the funding: £60 pounds had been allocated, but only £15 pounds was spent. It is a general and well-known rule of budgets that unspent balances are not renewed, on the reasonable ground that they had not been found necessary. The Association applied the rule, and the allocation was reduced every year until it had fallen to £10 pounds by 1845. Perhaps Forbes had taken the helm at the wrong time, just when he was forced to earn his own living and devote most of his energies to his mixed bag of new jobs. It was not until 1850 that he presented the first major report of the Dredging Committee to the Association, and even then it was largely confined to the results of his own cruises. He may not have trusted the reports from the local committees, or, more likely, he had not had the leisure to collate them, despite the standard forms. Whatever the cause, the committee seemed to be winding down in the late 1840s, and ceased its operations by the early 1850s. Forbes had set it in motion, but he was unable to propel it forward, and in the years leading up to his death it became almost moribund.

This was not quite the end of the story, however. John Gwyn Jeffreys was a lawyer at Lincoln's Inn who was an enthusiastic but little-known amateur naturalist who rather abruptly became an influential figure when he took over the chair of the Dredging Committee in 1861 and reinvigorated its activities, especially the dredging organized by local committees. It was a short-lived resuscitation. Although dredging from small boats was a fashionable summer recreation for a while, it gradually dwindled away without the fire that Forbes had initially kindled. Funding became a problem: more funding was supplied and more was spent, but the gentlemen dredgers had to shoulder most of the expense of cruises and began to complain about the drain on their pockets, not to mention the unaccustomed labor of pulling heavy weights on wet ropes out at sea. The meetings of the general committee became intermittent and took place at successively longer intervals, until its last meeting was held in 1873, after which it disappears from the records of the Association.

The Dredging Committee had succeeded in directing the attention of naturalists away from the land and toward the fauna of the seabed below the tideline. It was, in a sense, the victim of its own success, because the richness of the offshore fauna inevitably led men to wonder what might be found in the deep sea, far from land. Finding out what animals lived under a mile of water could not be attempted by heaving on ropes in a small boat, however: extending the dredging program to deep water required mechanical winches deployed from naval vessels. The necessity for new technology shifted the systematic exploration of the ocean floor from the Dredging Committee of the British Association to a new committee formed by an even more august body, the Circumnavigation Committee of the Royal Society, and, inexorably, from amateur naturalists to professional scientists.

Two Cruises

When Forbes set out in September 1854 on his last excursion, to North Berwick in the Firth of Forth, among his companions was a young man called Charles Wyville Thomson. They had met before: Wyville Thomson chaired the local committee that organized dredging on the east coast of Scotland. Like Forbes, he studied medicine at Edinburgh—unlike Forbes, he graduated MD—but, like Forbes again, he turned toward natural history, and indeed in 1870 took over Forbes's Regius chair in Edinburgh. Let us leave him there for the time being, while we look at a figure buried even deeper in the sands of time, William Benjamin Carpenter.

Carpenter was the eldest son of Lant Carpenter, a Unitarian minister in Bristol. It was a remarkable family. His elder sister was Mary Carpenter, who campaigned for the improved education of poor children and women. She was especially concerned with the problem of juvenile delinquents, and published a book with the unforgettable title of *Reformatory Schools for the Children of the Perishing and Dangerous Classes*. She was influential in drafting parliamentary legislation, and also traveled to India, where she formed the National Indian Association to improve conditions in schools, hospitals, and jails. William followed a more conventional path than his sister (although he too published a book with a memorable title, *The Use and Abuse of Alcoholic Liquors in Health and Disease*) eventually graduating MD—this is becoming a little repetitive—at Edinburgh in 1839. He became a prominent physiologist, and was elected Fellow of the Royal Society quite early in his career. There is nothing out of the ordinary in this. Two things came together, however, quite mundane in themselves, which together produced an effect altogether out of the ordinary. The first was that he developed a taste for natural history—this really is becoming a little repetitive—and even applied for Forbes's chair at Edinburgh, which he was refused on the sufficient grounds that he was a Unitarian. The second was his appointment as Registrar of the University of London in 1856. The one gave him the means and the time to pursue the other, where his interest lay, as it happened, in foraminiferans, of which much more later. For the present, they are minute, single-celled organisms living in the upper waters of the ocean, which meant that Carpenter, like Forbes and Wyville Thomson, had an interest in marine biology. Indeed, in the summer of 1855 he dredged crinoids (relatives of starfish, of which, again, more later) off the Isle of Arran—which brought him into direct contact with Wyville Thomson.

By the 1860s, Carpenter had become a vice President of the Royal Society and was in a position to exert a good deal of influence. He used it. Prompted (probably) by Wyville Thomson, he wrote a letter to the Council of the Royal Society recommending a much more ambitious undertaking than Forbes had dreamed of. It runs in part:

> Now as there are understood to be at the present time an unusual number of gun-boats and other cruisers on our northern and western coasts, which will probably remain on their stations until the end of the season, it has occurred to Prof. Wyville Thomson and myself, that the Admiralty, if moved thereto by the Council of the Royal Society, might be induced

to place one of these vessels at the disposal of ourselves and of any other Naturalists who might be willing to accompany us, for the purpose of carrying on a systematic course of deep-sea dredging for a month or six weeks of the present summer, commencing early in August.

The present summer, indeed. The Council consented; and the Secretary of the Society, William Sharpey, duly wrote to the Secretary of the Admiralty, Lord Henry Lennox, on 22 June 1868 as follows.

MY LORD, I am directed to acquaint you, for the information of the Lords Commissioners of the Admiralty, that the President and Council of the Royal Society have had under their consideration a proposal by Dr. Carpenter, Vice- President of the Royal Society, and Dr. Wyville Thomson, Professor of Natural History in Queen's College, Belfast, for conducting dredging operations at greater depths than have heretofore been attempted in the localities in which they desire to explore—the main purpose of such researches being to obtain information as to the existence, mode of life, and zoological relations of marine animals living at great depths, with a view to the solution of various questions relating to animal life, and having an important bearing on Geology and Palaeontology. The objects of the operations which they wish to undertake, and the course which they would propose to follow, as well as the aid they desire to obtain from the Admiralty, are more fully set forth in the letter of Dr. Carpenter to the President, and that of Professor Thomson, copies of which I herewith inclose. The President and Council are of opinion that important advantages may be expected to accrue to science from the proposed undertaking; accordingly they strongly recommend it to the favourable consideration of Her Majesty's Government, and earnestly hope that the Lords Commissioners of the Admiralty may be disposed to grant the aid requested. In such case the scientific appliances required would be provided for from funds at the disposal of the Royal Society.

Let me paraphrase this: a couple of our members, scientific gents, would like to find out what lives at the bottom of the sea, and would you please lend us a ship to help them find out? There was only one possible answer, which was, Yes.

It might have helped that the President of the Royal Society at that time, Sir Edward Sabine, had been one of the Scientific Advisers to the Admiralty appointed when the old Board of Longitude was abolished earlier in the century, and had since been active in matters of concern to the Navy, such as the first magnetic survey of the world. At all events, the Admiralty gave them a ship, the horrible old HMS *Lightning*.

Lightning was a wooden-hulled paddle gunboat with a displacement of 349 tons and a crew of 20. She was one of the first steam-powered vessels of the navy, designed and built in the uncertain age that separated sail from steam, with a paddlewheel mounted quite far forward, near the foremast. She was intended for survey work, with a shallow hull that was suited for shallow (and calm) water but was less than ideal for the boisterous waters of the northern seas. In the event, she set off from Stornaway late in the season, 11 August 1868, with Carpenter and Wyville Thomson on board, and returned, somewhat battered by gales, after less than a month, having had only nine days for dredging, and only four of these in deep water. These slender results were quite decisive, however: animal life flourished in deep water: starfish, brittle stars, crinoids, sponges, mollusks, and crustaceans were all recovered from 500 fathoms. This brief cruise showed that animals lived at a great depth, where the water was very cold and the seabed consisted largely of the chalky remains of foraminiferans. Above all, it showed that dredging in deep water is perfectly feasible, given a source of power and a stable platform.

The *Lightning* showed promise, but it also led ineluctably to the next mission, into even deeper water. Carpenter wrote a very enthusiastic report that reads like a grant proposal crafted to persuade a reluctant sponsor to disgorge, which of course it was. He concludes, "It is also greatly to be desired that these inquiries should be prosecuted at still greater depths." And so it was that the Council of the Royal Society, at its meeting on 21 January 1869, recommended that the Government should support a second and more extensive cruise. And so they did.

HMS *Porcupine* was also a wooden-hulled paddle steamer, a little larger than *Lightning*, and from all accounts much more comfortable. The summer was complicated, with three cruises of different lengths successively under the scientific direction of John Gwyn Jeffreys, Wyville Thomson, and Carpenter. The results were simple: when the dredge was worked for seven hours at a depth of 2,435 fathoms, in the northern part of the Bay of Biscay, it came up (thanks to the 12 hp donkey engine) loaded with animals— mollusks, echinoderms, crustaceans, annelids, sponges, and anemones. Life flourished in deep water. At the same time, experience began to improve the technology of deep-sea dredging. A heavy dredge lowered from a ship tends to plunge into the sediment and quickly fills with mud that impedes the capture of more organisms; the commander of the *Porcupine*, Captain Culver, suggested putting heavy weights on the cable a few hundred fathoms ahead of the dredge, so that it skims along the surface of the mud capturing any animal in its path. With simple modifications like this, deep-sea dredging began to develop from grab-and-haul to a more effective method of sampling animals from the deep sea.

The three cruises of the *Porcupine* in the summer of 1869 marked the end of the age of ignorance. It was now known beyond reasonable doubt that there were animals, lots of them, of many different kinds, in the deep sea. It was the conclusive refutation of Forbes's hypothesis of the lifeless depths, and to this extent it closed a question. At the same time, it opened a new and fertile field for enquiry, an entire new zoology as extensive as, or more extensive than, the old familiar zoology of the land and inland waters. The *Porcupine* voyages were not the end of the affair, but instead the beginning of the exploration of the deep sea. Carpenter still had unfinished business. In the autumn of 1871 he again wrote to the Admiralty, this time asking for a favor of quite staggering proportions.

The Circumnavigation Committee

There was, in fact, a lengthy intermission between his return from the *Porcupine* cruise and his final approach to the Admiralty during which Carpenter matured his plans. He was in an excellent position to exert influence at the highest levels. Besides his Royal Society position, he also belonged to the Metaphysical Society, whose members included the Prime Minister, Gladstone, and the Chancellor of the Exchequer, Robert Lowe. It may have been conversations with these contacts that led him to advise Sabine in November 1869 that Her Majesty's Government would support any reasonable request from the "scientific authorities." Nevertheless, for the next 18 months not very much moved forward; even the Royal Society Council was mute. It is possible that Carpenter was preparing the ground and developing his contacts; it is also conceivable that other events, such as the Franco-Prussian War, might have removed dredging to the distant outskirts of the political consciousness. The project began to move forward again early in 1871. Childers stepped down as First Lord, and his successor, George Goschen, now became Carpenter's target. At the June meeting of Council,

Carpenter suggested approaching the Admiralty for funding of a much more ambitious dredging voyage than any yet undertaken. Council listened; and deferred a decision until the autumn meeting. Carpenter did not wait for this. He must have petitioned the First Lord either in person or (much more likely) through the Hydrographer, George Henry Richards. At all events, when Council next met, in October 1871, he laid on the table a letter from Goschen promising support for an expedition, whenever the Society should choose to apply for it.

What persuaded the most powerful maritime force in the world to turn aside from its main business to detach a ship to dredge the bottom of the sea? Richards is mentioned as a matter of course in all the histories of the voyage as the channel between the Royal Society and the Admiralty, but I am inclined to think that he was more important than just a channel, important enough to make it worthwhile to describe him in more detail. In the first place, he was Navy to the bone. He entered the service as a midshipman aged 13, and stayed until he retired at the usual age of 55. He saw action in China, during the first Opium War, and in South America, at the storming of the Parana forts during a brief altercation with the Republic of Buenos Aires (now Argentina). Promoted Commander, he surveyed the coast of New Zealand and was later second-in-command of Belcher's fruitless attempt to find the Franklin expedition, making a remarkable 2,000-mile journey by sledge around Melville Island in northern Canada. He was promoted Captain when he returned to England, and soon set off again, this time to work on the US-Canada Boundary Commission to settle the maritime border through the San Juan Islands. He was not successful—the Americans then as now obstinately refusing to concede an inch of territory—but did survey the coast of Vancouver Island and the adjacent mainland. It was presumably the combination of courage and fortitude with punctilious attention to detail in surveying work that led to his appointment as Hydrographer to the Admiralty in 1864.

He was at this time (to judge from portraits) a full-fleshed man with an aquiline nose and magnificent mutton-chop whiskers. I may be imagining a slightly humorous curve to the mouth and I am certainly imagining a twinkle in the eye, but nevertheless it is a remarkable fact that everyone seemed to have liked George Richards. I cannot trace a word of criticism or dislike. During his survey of mainland British Columbia he insisted that the local place-names should be retained whenever possible, which has brought him grudging respect even from modern academic historians, an unparalleled achievement for a nineteenth-century British naval officer. There was no one else who could move as an equal both among the Fellows of the Royal Society and among the veteran sailors of the Admiralty, welcome in both worlds.

Above all, Richards was a thoroughly up-to-date hydrographer who was very well aware of the latest technological developments, especially in submarine telegraphy. The Atlantic Telegraph Company had been formed recently, in 1856, and succeeded in laying a cable in 1858 that worked for about three weeks. Repeated attempts to recover or relay the cable failed until a permanent connection was finally made in 1866. Richards became Hydrographer during the period in which the technical difficulties of cable-laying at depth were being encountered and overcome. Some of the problems were overcome by improving the cable itself, by using better insulating material, for example. The overriding problem, though, was that once the cable had been lowered into deep water no one knew much of the conditions it encountered: the depth, temperature, salinity, and pressure of the water column, and the nature of the sea floor itself. This could not be overcome without a systematic exploration of the deep sea. Richards clearly understood that a worldwide survey of the deep seabed was a strategic priority for the British Empire, and he used the prestige of the Royal Society as a lever to accomplish it. It was Richards, after all, and not the scientists, who laid out the course that *Challenger* would take. The advancement of knowledge in marine biology was a useful inducement, but it was not the *point*. The point, for the Hydrographer, was the hydrography, and the detachment of an outmoded vessel that could never stand in the line of battle was a cheap way to get the job done. The insistence of the academician Carpenter and the leverage of the hydrographer Richards proved to be the combination that turned a remote and improbable aspiration into a practical scheme that could be put into effect almost immediately.

The letter from Goschen was the turning point: a definite promise to fund a scientific expedition of unparalleled extent, from the only body capable of providing support for it, to the only body capable of conducting it. The amount of money involved was quite substantial. The cost of the cruise was estimated to be £23,470 per year, although £12,440 of this was the cost for the food and wages of the crew, which would have to be paid in any case. Altogether, the marginal cost of a scientific cruise lasting three and a half years came out to £40,025, or about £4.6 million (US$6 million) in current value. The agreement between the Royal Society and the Admiralty marked a decisive turning point in how scientific research is conducted.

Previous dredging expeditions (before *Lightning*) had been the responsibilities of gentlemen naturalists, who might like the cost to be defrayed from the funds of the Association—to which they paid a subscription—but would otherwise, grumbling, foot the bill themselves. The gigantic enterprise now projected was far beyond any private purse, at any rate any purse interested in dredging. Government support had for the first time made it possible to pursue scientific research on a scale hitherto unforeseen and unimaginable, except perhaps to Carpenter. The other side of the coin was that scientific research would now increasingly come under the control of government officials, who would, in time, determine its priorities and objectives, approve its facilities, and regulate its operations.

From this point on the expedition was a reality that could be derailed only by the defalcation of the Admiralty or the incompetence of the Royal Society. Neither failed. The Society sprang into action, as academies do, by forming a committee, at first clumsily named the Circumnavigation Dredging Committee, which later metamorphosed into the simpler Circumnavigation Committee. It was a very distinguished committee indeed: headed by the officers of the Society, it also included the principal dredgers, Carpenter, Wyville Thomson, and Gwyn Jeffreys; the Hydrographer to the Admiralty, Captain George Richards; and a clutch of eminent scientists, including Joseph Hooker, Thomas Huxley, and Sir William Thomson, with Alfred Russel Wallace added later. This was indeed one of the great committees of history. Its first action, on 10 November 1871, was to instruct Carpenter and Richards to draw up the general specifications for the voyage, which were approved by the Council later in the month and sent to the Admiralty in December. The voyage would explore the physical conditions of the deep sea; measure the chemical composition of seawater; determine the nature of the bottom deposits; and (finally) investigate the distribution of marine organisms. Throughout the world.

In past expeditions, two ships had commonly been required, for mutual support (*Erebus* and *Terror*, for example), but advances in ship design and construction made it possible to use a single ship, even for a long circumnavigation. The choice lay between *Challenger* and *Clio*, both laid up at Sheerness. The Naval Secretary, Robert Hall, suggested that *Challenger* was the more suitable, and it was so decided before the end of December 1871. The expedition now had funding, a purpose, and a ship. At this point, two crucial appointments had to be made before detailed preparations could begin: someone to steer the ship, and someone to steer the science.

The Captain was, of course, appointed by the Admiralty, who had a long list of possible candidates; but one stood out among all the rest. George Strong Nares was born to the sea, passing out second from the Royal Naval College in 1832. Had he passed out first, he would have been offered a command immediately, but as it was he had to wait. He was waiting one afternoon, as many hopeful or desperate officers did, in an anteroom at the Admiralty, hoping for an interview, and was about to leave, disappointed, when he encountered Richards, then a Commander. Richards offered him a position on Belcher's voyage to the Arctic, which he promptly accepted; an impulsive decision that determined the rest of his career. He worked his way up the ranks, from Mate, and in 1865 was appointed Captain of the paddle steamer *Salamander* to survey the coast of Queensland. He next moved to command of the *Shearwater*, on which Carpenter sailed in the cruise of 1871. In this way he came under the notice of both Carpenter, leading the dredging party at the Royal Society, and Richards, now the Hydrographer to the Navy. With these irresistible connections, arising from a chance encounter in a corridor, he was duly appointed in charge of the *Challenger* expedition.

One more thing, which might or might not be relevant. When the Suez Canal was due to be opened, Nares was in command of HMS *Newport* in the Mediterranean. The first ship through the newly opened Canal was to be the French imperial yacht *L'Aigle*, as was only fitting. Nares, however, somehow threaded his way in the darkness of the night of 16 November 1869 through the press of ships awaiting the opening, showing no lights, and was quite unexpectedly at the head of the queue, and ahead of *L'Aigle*, at first light. It was impossible to pass the *Newport*. The French were furious and, naturally, Nares was officially admonished by the Admiralty. Still, it had been a very fine piece of seamanship, and perhaps the Admiralty were not so displeased after all. They probably guffawed up their collective sleeve.

The choice of a scientific director was even easier, or so it seemed. Carpenter was in the entrails of the Royal Society, he had made all the political connections, and he had brokered the deal that made the expedition possible—and he wanted the job. He didn't get it, though. At its meeting on 4 April 1872 the Circumnavigation Committee (with Carpenter assenting) offered the position to Wyville Thomson, who accepted it, with minimal conditions, a week later. It would be Wyville Thomson, and not Carpenter, who would assume scientific direction of the voyage and garner most of the credit. Why was Carpenter not appointed? The usual reason given is his age; he was 58 at the time, against Wyville Thomson at 42, about the same age as Nares. Carpenter had decided, or been persuaded, that he did not really want to spend his sixtieth birthday in some distant port or sea. I am not so sure. Eminent as he was, he was not a very clubbable sort of man, and even his obituary refers to his "rather cold and formal" manner. Moreover, he also had a reputation as someone intolerant of criticism who picked fights and made enemies, defending his position with personal attacks. The *Eozoon* affair is a good example. Sir William Dawson, the distinguished Canadian geologist (and Principal of McGill College, as it then was) found a series of strange fossils in the ancient rocks of the Canadian Shield that he called *Eozoon*, the "dawn animal," and interpreted them as the remains of a primitive organism resembling a foraminiferan that was close to the ancestry of all living creatures. Carpenter was an expert on foraminiferans and agreed with Dawson's diagnosis, collaborated with Dawson in studying new specimens, and published several papers on *Eozoon*. Two much less prominent scientists, William King and Thomas Rowney of Queen's College Galway, came to a radically different conclusion, that *Eozoon* was not the remains of an animal at all, but merely a mineral deposited in minute layers. Carpenter was not amused by their presumption and made his anger very plain:

> That Professors King and Rowney, however, should take upon themselves to affirm that the structures which have been described by Dr. Dawson, Professor T. Rupert Jones, and myself, in *Eozoon Canadense* are not organic, would, I own, have greatly surprised me, if I had not had previous experience of the audacity (I use the word advisedly, and am quite prepared to justify it if called on) with which Professor King hazards denials of statements made by men who have a scientific reputation to lose, in regard to matters which they have carefully investigated. Until Professor King (I say nothing of his colleague, since it is scarcely to be supposed that a Professor of Chemistry should claim authority on Microscopic Palaeontology) shall have given some proof of his competence to estimate the value of evidence in this branch of scientific inquiry, I must rank him in the same category with those sagacious persons who still maintain that the flint implements were shaped out by a fortuitous succession of accidental blows, and not by human handiwork.

And so on, for several issues of *The Reader* in 1865. He made his anger too plain, in fact, because, regardless of the truth of the matter, everyone understood that personal attacks on colleagues who disagreed with you debased the currency of dispassionate scientific enquiry. Carpenter was simply not suitable to lead a long expedition in the course of which the ship's company, naval and civilian, were necessarily at close quarters for years on end. He would have been a disaster, falling out with the captain and setting the scientists at one another's throats. I think that the other members of the Circumnavigation Committee knew this—after all, they knew Carpenter very well—and someone, perhaps the Astronomer Royal, Sir George Airy, or Huxley, may have taken him aside for a quiet word. I can't prove this, but at all events the Circumnavigation Committee had made up its collective mind to entrust the leadership to Wyville Thomson.

4

The Ship and Her Crew

The Royal Navy is frugal in the naming of ships, and may recycle the same name for many generations of ships. The previous *Challenger*, launched in 1825, was a smaller wooden-hulled, fully-rigged sixth-rate (the smallest vessels rated as warships) armed with muzzle-loading broadside carronades. The next ship to bear the name, launched in 1902, was a much larger metal-hulled cruiser with engines generating 12,500 hp, armed with quick-firing breech-loading guns. Our *Challenger* was built in the uncertain transition period during which the sailing ships of the Napoleonic wars were gradually replaced by armored steamships. Consequently, she had both a full suit of sails and a steam engine, and looked rather strange, with the usual three masts but a funnel sticking up between foremast and mainmast. *Challenger* was a hybrid.

The Ship

Technically, *Challenger* was a corvette, the smallest of warships, a little smaller than a frigate. She was armed with 20 broadside cannon each capable of firing a 64-pound ball 3,000 yards with a muzzle velocity of over 1,500 feet per second, a highly reliable gun that would be a lethal weapon against any other small wooden ship that came within a mile of her. Even before her circumnavigation, *Challenger* had experience of blue-water sailing, serving on both the North American and Australian stations, and even acting for a while as the Australian flagship. For a dredging cruise, however, she required extensive modifications to make space for extra cabins and laboratories, apart from the specialized equipment used to operate the dredge. All her guns except two were removed, two 64-pounders being thought adequate to deter any casual pirates; in the event, they were never used during the cruise. Work began in the Sheerness dockyard in May under Richards's supervision, and the ship was ready for sea in November, all in the time that the Circumlocution Office might have taken to answer a letter (*Little Dorrit*

Full Fathom 5000. Graham Bell, Oxford University Press. © Graham Bell 2022.
DOI: 10.1093/oso/9780197541579.003.0005

Figure 1 HMS *Challenger* under full sail
Source: Vignette on, p. 60 of C. Wyville Thomson, *The Atlantic*, vol. 1.

had been published a few years before), and a striking testament to Richards's drive and efficiency in pursuit of an objective to which he had clearly given high priority.

The ship can be thought of as four layers stacked on top of one another, each serving a range of purposes. The upper deck, dominated by masts and sails, was open to the air and held the usual bridge and chart house, besides two whaleboats and a pinnace. It also held some less usual equipment. A pair of donkey engines, similar in design to a pumping engine, were bolted to the deck to the left (port side) of the mainmast; these were used for hauling in the dredge. Just in front of the mast, projecting above the ship's boats, was a most unusual structure, the dredging platform from which the naturalists could observe the operation of the dredge. There was also a small deckhouse aft for the use of the naturalists. A supply of rope for dredging and sounding was stored near the bows. Rope! The *Challenger* carried a vast amount of rope: 50,000 fathoms—300,000 feet—of rope, in all sizes up to three-inch diameter, enough rope to stretch from Sheerness to Portsmouth and back.

Yet more rope had to be added during the voyage. It was all needed because fouling the bottom, as often occurred, would often result in the loss of the dredge and a thousand fathoms or more of rope. It was all hemp rope, made of twisted plant fiber, and it worked well enough; the breaking strain of the two- to two-and-a-half-inch hemp rope that was usually used to tow the dredge was about two tons. Wire was unreliable, being difficult to coil and liable to kink, and the stranded steel wire that would have been much preferable was not available until a few years later. Hauling in the dredge from deep water, then, involved reeling in a couple of miles of thick wet rope. This was a long way from amateur dredging in shoal water with window-cord.

The main deck immediately below had been extensively modified for a scientific cruise. By tradition, the aft section of the main deck was reserved for the great cabin of the captain, who ate and slept there, more or less in solitary confinement, in isolation from the rest of the crew. This reflected the very strict hierarchy of ranks, usual in any effective military force, that determines who may give orders to whom. In the *Challenger* this practice had to be waived to some extent because of the coequal status, in some respects, of Captain and Director. (Wyville Thomson had earlier requested a nominal temporary rank and uniform for the scientific staff, but this was inevitably turned down by the Admiralty because it implied that civilians might give orders to seamen, enforceable under naval discipline, which was clearly absurd.) The sleeping cabin and fore cabin (bedroom and sitting room) were thus shared between Nares and Wyville Thomson. The zoological laboratory adjoined this space on the port side. It was an attractive and well-lit space with skylights and portholes, fitted with cupboards, drawers, and racks to hold specimens and bottles, and a long table running down the center where the microscopes could be used. It looks in the pictures like the very paradise of a zoologist, and so it may have been in a flat calm on a bright morning. With the ship working in even a moderate sea it was not as perfect; for one thing, all microscopes at that time used light reflected from a mirror through the stage, and with the light shifting as the ship moved, it was often not possible to view a specimen for more than a few seconds at a time. Nevertheless, it was the very first well-equipped zoological laboratory on board a ship, and served its purpose well enough. Opposite was the chart room for the surveying officers, and further forward the chemistry laboratory and the photographer's workroom. Beyond this lay, as usual, the main cooking range, the seamen's lavatory, and the sickroom—besides, of course, yet more storage for yet more rope.

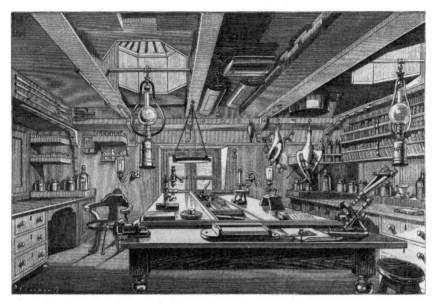

Figure 2 The zoological laboratory on board Challenger. This is where Wyville Thomson, Murray, Moseley, and Willemoes-Suhm worked.
Source: Narrative, vol. 1, Part 1, p. 6.

The lower deck had been modified by breaking out a series of small cabins for the scientific staff, as well as for the officers and petty officers of the crew. These were largely grouped around the wardroom, where the officers and scientific staff ate and (ideally) socialized; the petty officers had their own mess a little further forward. The ship's company, the able-bodied seamen who formed the greater part of the crew, ate and slept in the forward part of the lower deck. The arrangement of space reflected the hierarchy of rank, the more senior occupying the more spacious quarters where some privacy was possible, while the majority huddled together in a confined space where privacy was unobtainable. It is an arrangement that preserves order by separating interests, and thereby enhances the efficiency of the ship as a whole, or aids in the subordination of the greater number, according to the point of view, and if this seems a typically Victorian instrument of oppression, it is also an arrangement that serves the same purposes in any large town nowadays, and if you doubt this, look around you.

Below the lower deck lay the hold, below which there was only seawater beyond a few inches of wood and copper. The hold in most ships of the time held only the stores of food, water, and spirits (this last, guarded by marines).

In the *Challenger* it also held the engine house, boilers, and coal stores. It was a dark and fetid space, occupied chiefly by engineers and rats. The engine itself was a two-cylinder steam engine developing 400 hp that turned a single screw that could be folded out of the way when not in use. It was not used for cruising, unless the wind failed completely, because it was too feeble (a World War II destroyer, 80 years later, would have 50,000 hp) and the *Challenger* did not carry enough coal for a long voyage under power. It was essential, however, in keeping the ship head to wind during dredging and sounding. It was also useful for maneuvering in confined spaces, and on one tumultuous night in the Antarctic the engine rescued the ship from a perilous situation that might otherwise have brought the voyage to an abrupt and disastrous conclusion.

The Scientists

After the appointment of Nares as Captain and Wyville Thomson as Director, the crew, naval and civilian, was assembled from the top down. I doubt that so heterogeneous a collection of men had ever before been assembled to work together in a space so confined. Wyville Thomson had more or less a free hand, or at least an overriding veto, in the nomination of the scientists, and asked the Circumnavigation Committee for a staff of four: a secretary-cum-artist, two naturalists, and a chemist. He drew up a list within a few weeks, presented it to the committee, and Council approved it on 20 June.

The chemist was to be John Young Buchanan, the laboratory assistant to Professor Alexander Crum Brown at Edinburgh. The naturalists were Henry Moseley, William Stirling, and John Murray. Moseley was an Oxford graduate, at this point on a traveling fellowship; Stirling was the Falconer fellow at Edinburgh; and John Murray had just completed a geology degree under Professor Archibald Geikie at Edinburgh. John James Wild was appointed as artist to the expedition, doubling as secretary to the Director.

The list has two rather odd features. In the first place, being at Edinburgh is clearly an advantage: Buchanan, Stirling, and Murray are all associated with the University in one way or another. Thomson knew Wild from Belfast, so Moseley was the only English outsider on the list. The second odd feature is that there are now three naturalists, rather than the two positions that had been agreed. This led to a long argument with the Treasury, which eventually agreed to the increase in salary mass only if this could be made up by a corresponding reduction in supplies. The Admiralty agreed, but sub-sequently clawed back £100 for feeding the extra naturalist; in short, the usual interdepartmental wrangle, one in which the Admiralty for once had the better of the Treasury. It wouldn't happen nowadays. In fact, Wyville Thomson could have avoided the argument, because in late September, just before the scientific staff were due to be formally appointed, Stirling was dropped. It is not quite clear whether he resigned (most printed sources) or was dismissed "for certain reasons" (as Wyville Thomson put it in a letter to Huxley). At all events, the naturalists were now down to two, but Wyville Thomson was determined to hang on to three, because he promptly ap-pointed Rudolf von Willemoes-Suhm, a young German naturalist whom he had just met at, ahem, Edinburgh.

It was an uneven list. Buchanan was thoroughly competent and was re-sponsible for the impeccable and extremely detailed physical chemistry of seawater that was pursued throughout the voyage. Wild was painstaking enough and could take and develop photographs, not a common skill at that time. Moseley was a mistake, I think. He was appointed as the botanist to the expedition and worked mainly during shore excursions, since there are no plants in the deep sea. At the very beginning of the book he wrote after the voyage, he remarks, "Since the results obtained by deep-sea dredging, even in most widely distant localities, were very similar and somewhat monotonous, all reference to them will be deferred to the end of this narrative." This is not the spirit at all. He could at least have taken an interest in the planktonic algae. Willemoes-Suhm was a gamble made on the spur of the moment to make up for the premature departure of Stirling. John Murray, on the other hand, was a superb appointment. He was the oldest of them (at 31) and the only one who had actually been at sea—he sailed as surgeon to a whaling ship on a voyage to Spitzbergen and Jan Mayen Island in 1868. He turned out to be a dogged, indefatigable character who when the voyage was over took charge of assembling and publishing the long series of Reports that secured the lasting reputation of the expedition and laid the foundation for the future development of biological oceanography.

These were the men who sailed together to find out what lives in the depths of the sea. How, and why, had they been chosen? It is almost as though Wyville Thomson had signed up the first students he met strolling along a university corridor in Edinburgh. After all the decades of increasingly am-bitious dredging, after involving the most eminent scientists in the country, after striking a unique bargain with the Admiralty, after all this, one might have expected a fairly lengthy consultation with senior colleagues to arrive at a list of eminent and experienced specialists. Nothing of the sort happened. None of those chosen had an established reputation, and only one had even been to sea before. They did have one thing in common: all were young, mostly in their twenties. This might be no bad thing: it meant that none had strong family ties and all were healthy and vigorous, fit for arduous work at close quarters during a long voyage. It also meant that all were distinctly junior to Wyville Thomson and deeply in his debt for the opportunity he had provided for them. It was going to be his expedition.

The Crew

Dredging in shallow water from a small boat needs someone with one hand on the tiller, and another (at least) to handle the sail, while the dredge is put overboard and recovered. It can all be done turn and turn about. Dredging from a ship in deep water is a very different proposition. A ship is a floating village, with the added complication that it must be made to float in a particular direction, without running out of supplies, being beaten off course, or striking a reef. All of this requires a large crew with a very detailed division of labor overseen by peremptory authority. It cannot be done by committee and it certainly cannot be done by naturalists. In all the published accounts the naturalists take a prominent position because they were, after all, the first cause and prime movers of the expedition, or so they thought, and have always claimed it for their own. It requires a little effort to recall that there were only six scientific staff among about 200 crew, and that none of the former were permitted to give any order to any of the latter (cabin servants of course excepted). This was emphatically a naval cruise that had been agreed, implemented, and organized by Captain George Richards, the Hydrographer to the Admiralty.

Once the ship's complement had been approved (in late September), it was up to Richards to find the men. He went about this in a quite extraordinary and unprecedented manner, by sending out a call throughout the Royal Navy for volunteers. It is not normal, then or now, to call for volunteers; sailors are drafted to a ship and that is that. Some of the officers were, indeed, transferred in the usual manner. In particular, Thomas Tizard had been the navigating officer of *Shearwater* when Nares had commanded her on the Mediterranean cruise in 1871, with Carpenter on board, and was named as Navigator to *Challenger* in April 1872, I presume on Nares's recommendation. For most of the rest, however, Richards had the pick. It was a curious reversal of the normal procedure, or at least what we think of as the normal procedure, in which scientific appointments are made by open competition whereas military appointments are made by rank and seniority. Richards, in contrast with Wyville Thomson, used open competition—and then, of course, chose among those who put themselves forward. It was going to be his expedition.

There were 21 officers assigned to *Challenger*. Captain, Commander (the second-in-command, J. L. P. Maclear, chosen by Richards), lieutenants, paymaster (the purser), surgeon, chief engineer, and their assistants. There was also the full range of warrant officers, such as boatswain, carpenter, and steward; no chaplain, though, as the ship was a little too small. There were also about 200 crew to man the ship, including marines; it depends how you count the junior petty officers and their mates. Fifty of the crew that Richards chose were Boys First Class, which seems like an awful lot of boys, even if they were first class. I have (with a little help from my wife) raised three boys, all first class; but 50 might be more than a handful. A schoolmaster was shipped—Adam Ebbels—and no doubt earned his pay.

Figure 3 Officers and scientists of the *Challenger*. The photograph was taken in October 1874, probably at Ternate. The structure overhead is the dredging platform.

Source: Album 2 of the assistant paymaster, John Hynes, ALB0175 in the Caird Library of the National Maritime Museum, Greenwich.

The officers are all known; most of the seamen and boys are unknown. History is written, not by the conquerors, nor by the conquered, but by the literate; seamen were almost all illiterate. There were several books written about the voyage, afterward, by the naturalists or the officers, as usual, and none by the seamen, as usual. Except that this was not quite the case on this voyage. One of the boys was Joseph Matkin, Ship's Steward's Boy, who joined the ship on 2 December. He was literate, being the son of a well-to-do printer and stationer in Rutland, and even verbose; his surviving letters home fill a book. His position was a little sensitive, being among the seamen but not one of them, because he could not be called on deck (except in extreme emergency) but was responsible, under the Steward, for the distribution of food and drink, which might have put him in the way of awkward (or profitable) pressure, except that for this reason he messed with the petty officers. His voice is not the authentic unheard voice of the lower deck, but it is a near as we are likely to hear. His letters home are, so to speak, the unauthorized version of the voyage.

Several other members of the crew wrote up their account of the voyage afterward. The official version is the *Report* of the voyage, which was mostly Murray's work; the first part of this is the *Narrative* of the cruise, written by Tizard, Moseley, Buchanan, and Murray, and published in 1885. The *Summary* published by Murray in 1895 is a useful station-by-station travelogue. Wyville Thomson published an account of the Atlantic sections of the cruise in 1877, the same year that Engineer W. J. J. Spry published a detailed description of the entire cruise. Two rather obscure books about the voyage appeared at about the same time, the *Log-Letters* of Sublieutenant Lord George Campbell in 1877, and *At Anchor* by the artist J. J. Wild in 1878. The letters written by Willemoes-Suhm from the *Challenger* were published posthumously in 1877. Herbert Swire, a navigating sublieutenant, left it until 1937 to publish his book. The voyage was, then, voluminously documented in thousands of pages written by the men actually on board, and there is no difficulty in following it from day to day, often in minute detail. Having said that, the accounts are often redundant, tend to emphasize the shore visits rather than the open-water cruising, omit or gloss over some important aspects of the voyage, and, with the exception of Matkin's letters, give only the main-deck point of view. Even a few pages by one of the Able Seamen who made up the bulk of the crew might provide a very different perspective, but those few pages were never written.

Toward the end of November, 243 scientists, officers, and men began packing their bags, saying their goodbyes, and moving by foot, train, or coach to the Sheerness dockyard, where *Challenger* lay. The crew began arriving on a Monday morning and walked across the gangplank to fill the ship to find their quarters. There were close on 200 of them, and none would see England again for nearly four years, after traveling around the world and crossing the equator six times. Some—nearly half—would never return, having died or deserted or fallen sick in some distant port. One of them—Tom Tubbs, a Marine—would not even begin the voyage. Returning to shore in the dark November evening, he missed the gangplank (no lights or safety rail then) and fell between boat and dock into 30 feet of cold water. A diver recovered his body the following day, and a replacement soon arrived from Chatham Barracks. No unfortunate death would delay this voyage. Two weeks later, on 7 December 1872, the ship left Sheerness and stood out for the Channel.

PART 2
THE CRUISE

5

Outward Bound

December 1872–February 1873

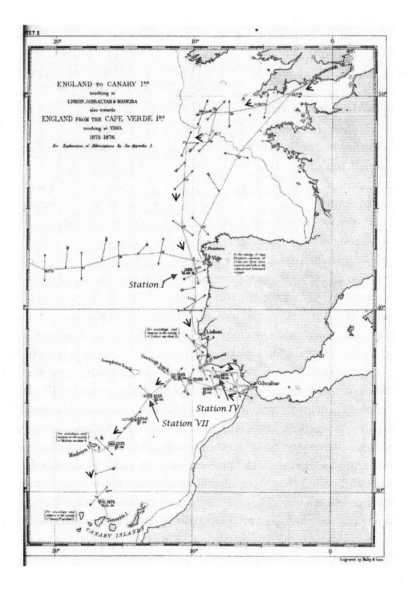

Full Fathom 5000. Graham Bell, Oxford University Press. © Graham Bell 2022.
DOI: 10.1093/oso/9780197541579.003.0006

Sheerness to Portsmouth

The first leg of the voyage might seem uneventful if you relied on the books, although Spry and Murray noted that the weather was stormy. Only Matkin was more forthright: Sunday was "the most fearful night I ever passed in my life." *Challenger* had sailed just in time to be confronted by a force 10 gale, just two points short of a hurricane, blowing from the west-southwest, almost directly contrary to her route. The windspeed rose to 50 knots and small ships were scattered all along the coast from Kent to Cornwall. The *Challenger* was too large to be in serious danger, but even so one of her boats was smashed by the waves and the jib boom carried away. Instead of a brisk day's sail to Portsmouth *Challenger* limped along the coast, putting in successively at Deal, Folkestone, and Dungeness before anchoring at Spithead on Thursday, having made about one mile per hour on the way.

The scientific staff were no longer on board when Portsmouth was finally reached. They had been prostrated with seasickness on that first tumultuous day and left the ship at Deal to take the train to Portsmouth, where they arrived, of course, two days before the *Challenger*. This cost them a certain amount of dignity; seasickness is always a little comical to those unaffected by it, and "the sailors did make game of them," according to Matkin. What the crew thought when the scientists abandoned ship and took to the train can only be imagined. It had been a disconcerting beginning to the voyage.

It was also an indication of the complex stresses that are bound to develop on a survey ship. On a regular naval voyage there is inevitably some degree of tension between officers and men, which the navy traditionally manages by strict social segregation, a rigid hierarchy, and instant obedience to orders given by a superior. On a survey ship (then as now) there are two further sources of tension, not only between scientists and crew, but also on occasion between scientists and officers. It is not always convenient to lower the dredge, and on many occasions the dredge was not lowered because it did not suit the management of the ship. Soundings were different; all ships took soundings, and even on *Challenger* the number of soundings far exceeded the number of dredges. In the main, however, scientists and officers rubbed along well enough because they were all middle class (except George Campbell) and more or less at a level. They were paid more or less the same, for example (the figures can be multiplied by about 100 to make them comparable to current values, although this is a rough guide at best). Wyville Thomson had negotiated £1,000 per year, perhaps less than his Edinburgh salary but still comfortable. Captain Nares, as captain, was paid less than half of this, but the special allowance for surveying duties would have brought his total salary up to about £775. Commander Maclear was paid about £365 and Tizard (again, with surveying allowances) perhaps £500 or so. The naturalists were paid £200 each, about the same as the lieutenants at £182. I do not know why Wild, the secretary/artist, was paid twice as much as the naturalists, nor whether the naturalists knew of this. Still, in the main salaries were not very different between the scientific staff and their rough equivalents among the naval officers.

Below this top tier, however, there was a wide range of salaries. The Chief Engineer would have been paid almost as much as the captain, some £400 per year, although with no surveying supplement. The Engineers and Surgeon would earn about £250 and the Warrant Officers—boatswain, carpenter, and so forth—about £150. The steward, as a chief petty officer, would have £57 at the most, and an able seaman about half of this. As ship's steward's boy, Matkin initially earned seven pence per day—although this was doubled when he was promoted a year later—or about 1 percent of Wyville Thomson's salary. In other words, the scientific staff collectively earned about as much as all the able seamen combined. Extreme inequality of income is not exclusively a modern phenomenon, and it produced, as it must always produce, a certain level of resentment, not always voiced. Especially when these overpaid seasick gentlemen take the train.

Portsmouth to Lisbon

Challenger completed her fitting-out in Portsmouth and sailed for the Bay of Biscay shortly after noon on Saturday, 21 December 1872. You may imagine how the families of the crew felt; with the prospect of a voyage lasting more than three years, the ship sails a few days before Christmas, the feast above all that brings the family together. It would have done no harm to delay a week, and I do not know who took the decision (Nares was a married man), but revenge was not long in coming, in the shape of the great turkey mystery.

The *Challenger* sailed down-Channel on a headwind, but the seas were still very rough and Christmas was spent on a rolling, pitching ship with smashed pottery on the decks of the messes and even experienced sailors seasick. On the other hand, there was an issue of one-third of a pint of sherry to all hands on Christmas Day, with a meat pie and plum pudding for dinner, which sounds like just the thing, unless, of course, you happen to be seasick. The officers, naturally, had the full traditional Christmas dinner, including a goose as well as a roast turkey, which was just then coming into fashion. At least they would have done, except that turkey and goose both disappeared while the cook's back was turned, and nothing was ever seen of them again except for a few goose bones at the top of the mainmast. The officers were naturally disconcerted, but made no progress at all in their search for a culprit. They did not mention the incident, either, in their accounts of the voyage; my authority is Matkin, who also does not name a culprit, although he must have heard some rumor if any of the hands had been responsible. It all sounds a likely story. The cook's back is never turned on roast meat, and its absence would anyway have been noticed instantly. Nobody would take a large bird up the rigging in full view of the bridge. I imagine that it was a prank; some group of republicans, disappointed of their Christmas at home, decided that those responsible would be disappointed too. It would be interesting to know who was responsible, but it is now past all conjecture.

With the turkey digested by the party responsible, the crew returned to their normal diet. Breakfast at 6:00 a.m. was cocoa and hard biscuit, followed by the issue of lime juice at 10:30. Dinner at 11:30 was salt pork, salt beef, or tinned beef, with pease pudding, plum duff, or preserved potatoes. The rum ration was issued at noon: one-half cup of rum in one cup of water. Supper at 5:00 p.m. was tea and hard biscuit. Hard biscuit, indeed; it was a hard service.

Station I: 30 December 1872

41°58′ N, 9°42′ W; off Cape Finisterre, 1,125 fathoms

Challenger passed Ile d'Ouessant, off the western tip of Brittany, and struck off southwest across the mouth of the Bay of Biscay. Sailing ships try to avoid bays, semienclosed bodies of water between headlands, because an onshore wind turns a bay into a lethal trap. The Bay of Biscay was particularly notorious because the prevailing westerly winds can make it impossible for a square-rigged ship to weather either headland, pushing it inexorably toward the coast of France. With a steam engine in her belly, feeble though it was, *Challenger* had little cause for worry, but kept well wide of the Bay anyway. Nevertheless, the weather was stormy, and battering into the big waves, swept across a thousand miles of Atlantic Ocean, must have made the ship uncomfortable even for the cosseted naturalists. It did not moderate until the ship rounded Cape Finisterre, another place avoided by sailing ships because of its complex on-shore currents. If I had been in charge, I might have tried out the dredge and hauling gear first on a soft bottom in shoal water, but with the weather still boisterous this may not have been practicable, and anyway was not done. Nine days out from Portsmouth, the first sounding was taken 40 miles off Vigo.

Sounding and dredging in theory and practice

The principle of sounding is very simple: a heavy weight attached to a rope marked at intervals is lowered into the sea until the bottom is reached and the depth recorded. In practice it is a good deal more difficult because the forward motion of the ship tends to swing the sounding line into a diagonal that overestimates the depth; anyone who has fished in deep water will be familiar with the surprising difficulty of keeping the line anywhere near vertical from a moving boat, even with heavy sinkers. This difficulty had been foreseen, of course, and mitigated by sounding from a platform extending from the ship's side, near the bow, and by using the main engine to maintain station after the ship had been brought head to wind. The sounding lead itself was not merely a heavy lump of metal, but a fairly sophisticated device designed to bring up a sample of the seabed as well as recording the depth. At its simplest, it consisted of a heavy sinker cast around an iron rod terminating in a hollow tube about a foot long, with a pair of butterfly valves, opening inward, at its base. The valves opened during descent and admitted a sample of sediment when the bottom was reached, then closed when the lead was pulled up to retain the sample. The sounding line would also normally carry thermometers designed to resist very high pressure and devices for collecting a sample of seawater at the bottom and at intermediate depths.

At its first cast, the sounding line found 1,125 fathoms, but broke while being recovered, with the thermometer lost, which was why the *Challenger* had been so plentifully equipped with spares. The dredge was then lowered for the first time in a little over 1,000 fathoms of water. The dredge itself was a bag supported by a rectangular iron frame. The bag was inevitably a compromise: too coarse a mesh, and animals would be lost; too fine, and the bag, fouled by mud, would sweep aside any animals that it encountered. A bar mounting a row of swabs looking like mop-heads or cheap wigs was attached behind the bag. The swabs were added on the advice of Captain Calver of *Porcupine* and proved to be unexpectedly effective, especially for entangling spiny animals such as starfish and bushy animals such as bryozoans. With the ship brought before the wind and held (more or less) steady by the screw, lowering the dredge and its attached weights was not difficult. Recovering a ton or more of gear (two miles or so of rope, weights, and full dredge) from the sea bottom was quite another matter. Here is a highly simplified description of how the dredge was brought back to the surface. I have tried to avoid specialized nautical terms that I do not fully understand myself.

The apparatus used to recover the dredge consisted of two sets of ropes and pulleys. The first set is based on a short rope suspended from a pulley attached, by separate ropes, to the top of the mainmast and the end of the yardarm (the end of the horizontal spar attached to the mainmast, used for hanging the sail). This pendant rope bears the "accumulators," a rather Heath-Robinson cluster of thick elastic bands intended to accommodate any sudden strain. Beneath the accumulators is a short rope leading to a pulley through which the dredge rope passes.

The dredge rope is led through this pulley to a second pulley secured inboard to the forecastle, and from there to a drum on the port side. Three or four turns are taken around this drum, and the rope then led to a pulley on the port side of the deck, from there to a pulley on the starboard side, and from there to the starboard drum. The rope then passes from the starboard drum to the cord storage rack. The port and starboard drums are connected by a long crank axle, extending across the beam of the ship, which is turned via gearing by the piston rods of the two donkey engines, working in concert. Consequently, when the donkey engines are put into gear, the drums revolve and start pulling in the dredge. The dredge rope is first wound a few times around the port drum to prevent slippage, then wound in by the starboard drum, and eventually passed from this drum to storage.

All clear now? The point of this complex and ingenious arrangement is that the dredge rope is boomed out clear of the ship's side (from the yardarm), ensured against shock (by the accumulators), and protected from slipping when being hauled in (by the port drum). Once it has been recovered the dredge can be swung onto the dredge platform, above the upper deck, and discharged. It all sounds quite simple in theory, but the practice, on a heaving deck with the spray flying and the unprecedented weight and power of the gear, was very different. It was successful, though; up came the dredge from its first haul, but alas turned upside down, with nothing to show.

A soft starfish

About two o'clock in the afternoon the dredge was sent down again, left for an hour and then hauled in, reaching the dredging platform at 5:30 p.m. This time it came up full of blue mud, the sediment of silt and clay that accumulates in deep water from a nearby coast. You can buy it nowadays, if you wish, as a face cream, guaranteed to rejuvenate the skin. The naturalists, mostly juvenile anyway, grubbed in it with sleeves rolled up to their elbows, eager to explore the first prize snatched from the deep, the first systematically taken sample of the deep ocean floor. They found crustaceans, brittle stars, and a bright red starfish of an unknown kind.

Finding, and naming, new species had been one of the principal tasks of biologists for the last 100 years or more. The promise laid out by Linnaeus a

Figure 4 *Hymenaster* from Station I

Source: W. Percy Sladen (1889), Report on the Asteroidea. Zoology, Part LI, Plate LXXXVII. Bound in Reports, vol. 30.

century earlier of constructing a complete catalog of all created beings was for long the ultimate objective of biology, because it would reveal God's plan for the world. Taxonomy would be the key to understanding God's mind. This cosmic vision was wrecked when Darwin showed how new species could arise through natural laws, but in 1872 this idea was still controversial, and, besides, the glamour of seeing a new kind of animal for the first time, and attaching your name to it, in perpetuity, was a powerful draw. It was also a persuasive defense of the *Challenger* project, at least to an academic audience: we said that we would discover a new world, and here is its first representative. The red blob was accordingly given the name of *Hymenaster membranaceus* Wyville Thomson, 1873, following the standard system of zoological nomenclature.

It was a very odd animal. Starfish and their relatives are pretty odd anyway, but this was doubly odd. The iconic starfish of the seashore is a rather leathery rough animal with five clearly separated arms. *Hymenaster* has a gelatinous translucent body wall through which the internal structures are clearly visible, with a sheet of tissue, supported by lateral spines, extending between the arms. Living specimens have been captured by subsequent expeditions, and they can exude mucus by the bucketful, perhaps to deter predators, which is why they are called slime stars. They are by no means alone; many kinds of animal, including fish and mollusks, have gelatinous species in the deep sea. It is not clear why, but it may be that we are simply looking at the animals from the wrong point of view. We do not find it difficult to understand why animals should have tough bodies supported and protected by strong skeletons or shells because we think of them living in the zone of the sea that is most familiar to us, between and immediately below the tides. Animals living here experience strong physical forces such as wave impact, turbulence, and current shear, and they have evolved to resist them: a gelatinous animal would quickly disintegrate. At the bottom of the deep sea these forces do not act, or act much more weakly, and are therefore no longer effective agents of natural selection. Gelatinous bodies are cheap to make—they consist largely of interstitial fluid—and not very rewarding to eat, so they are likely to evolve repeatedly in different kinds of animal living in the deep sea. We find them strange only because we look at them from the wholly inappropriate point of view of a large terrestrial mammal.

At all events, the large terrestrial mammals on the dredging platform of the *Challenger* could be well satisfied. At the first successful cast of the dredge, the scientists had brought up a distinctive, hitherto unknown animal from the hidden depths of the sea. After all, this was going to work.

From here the *Challenger* sailed south off the coast of Portugal over hilly terrain just seaward of the shelf, across a field of blue mud. The sampling gear was lowered into deep water on 2 January, somewhere in the complicated series of canyons north of Lisbon, but the sounding rope broke and the dredge too was lost, with about 2,000 fathoms of line. The following day she moored in the Tagus, giving the crew a chance to visit Lisbon and the officers and scientists the excuse to entertain royalty. She spent the next 10 days at anchor, not merely for relaxation but also to take on coal, a recurring necessity throughout the voyage.

Station IV: 16 January 1873

36°25′ N, 8°12′ W; off Cape St. Vincent, 600 fathoms

The *Challenger* finally left Lisbon on 13 January (with some reluctance, I think) and sailed south over blue mud, green mud, green sand—deposits largely washed down from the land—with the soundings swinging erratically up and down as the ship passed over shelf and canyon. She crept slowly down the coast, sounding and dredging by day and sailing only at night, in light winds. Two days later she rounded Cape St. Vincent and turned east for Gibraltar and, unexpectedly, into a coral sea.

Gorgonians

When the dredge came up on 16 January it was full of animals, and had (in Murray's words) "great masses of a Gorgonoid entangled in the rope and about the mouth of the dredge." This was by no means as terrifying as it might sound: a gorgonian is a type of coral, quite harmless, although very different from the branched or bouldery stony reef corals familiar from a thousand nature documentaries. Reef corals are shallow-water animals that rely heavily on photosynthetic algae growing within their tissues. Far below, quite different kinds of coral grow in the cold dark depths. Among them are gorgonians, the sea whips, soft (that is, nonstony) corals related to the sea fans that can sometimes be seen swaying in the sandy valleys of coral reefs.

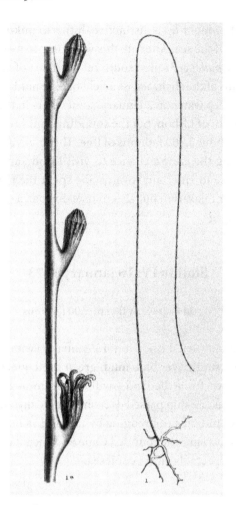

Figure 5 *Strophogorgia* from Station IV
Source: E. Perceval Wright (1889), Report on Alcyonaria. Zoology, Part LXIV, Plate 1. Bound in
Reports, vol. 31.

(Incidentally, nothing illustrates our ignorance of marine animals better
than our usual manner of naming them, which generally consists of nomi-
nating some familiar object to which they bear a superficial resemblance and
prefixing "sea," as in sea fan, sea whip, sea anemone, sea cucumber, sea biscuit, and so on; even seahorse and sea cow.)

The dredge had plowed through a patch of *Strophogorgia*, sea whips,
plastering their flexible black stems over its mouth. Although they are very
different from reef corals in their overall appearance, gorgonians have a

fundamentally similar body plan. Most corals are colonies that comprise hundreds or thousands of minute polyps, each resembling a miniature sea anemone. Each polyp has a ring of tentacles surrounding the mouth, which leads to a spacious chamber in which the prey are digested. There is no gut or anus—any undigested remnants are simply voided back through the mouth. The prey are small shrimp-like creatures such as copepods that are captured by the stinging cells borne on the tentacles. These are very unusual cells, each containing a poisoned harpoon attached to a long filament. When the cell is triggered by the movement of a small animal nearby, the harpoon is forcefully ejected at high speed. It hits the copepod with the force of an antitank missile, blasting through the cuticle and injecting it with a powerful neurotoxin. A copepod struck simultaneously by a few dozen nematocysts is killed instantly and tethered to the polyp by their filaments. The tentacles of the polyp now begin to bend, slowly cramming the dead prey into its mouth.

The strangest feature of these corals, from our point of view, is their dual nature, as polyps and as colonies. Each polyp is in principle complete in itself, with autonomous feeding and digestive structures, nerves, muscle fibers, and gonads. It looks as though it could be simply removed from the colony and live comfortably by itself, and there are in fact plenty of species of solitary corals that do just that. On the other hand, all the polyps of the colony are embedded in a common ground substance containing wandering cells that do not belong to any polyp in particular. This ground substance also contains nerve fibers that hook up the nervous systems of neighboring polyps and enable them to communicate with one another. Moreover, polyps can exchange nutrients because the digestive chambers of the polyps are connected with one another by channels or strands of tissue, so that the prey that was captured and is now being digested by one polyp can be shared among its neighbors.

Many kinds of animal have independently evolved colonial organization and consist of partially connected zooids (the more general term for polyps); we shall encounter bryozoans and tunicates, for example, later in the voyage. To have some sense of how they live, imagine that you are sitting in a dark, crowded theater. The walls of your seat enclose you closely and completely, except that the top is open, or has a lid, so that you can stick your hands out and grab any piece of popcorn that floats by. On both sides of your seat there is a hole in the wall through which your skin passes, forming a tube that connects you with the body cavity of your neighbor on either side. There is also some kind of direct or indirect sensory connection, so that you are aware, rather dimly, of your neighbor's state of mind, or at least their state of health. If there is an alarm of danger, you may be able, if you are a tunicate, to quickly disengage and dash for the exit; or you may just have to sit it out, if you are a coral or a bryozoan. Otherwise, you are there for the duration of the performance, which lasts a lifetime.

What, then, is the individual—is it the polyp, or the colony? This is a purely philosophical question (that is, one incapable of resolution) because it arises from our habit of seeing the world exclusively from our own point of view. We are strongly individualized animals and assume that this is the rule, whereas any other kind of organization is odd or exceptional, like conjoined twins. It is certainly true that many familiar marine animals—fish, crabs, whelks—are unmistakably individual, but many of the less familiar groups—corals, hydrozoans, bryozoans, ectoprocts, phoronids, tunicates—are partly or entirely colonial. In animals like these there is no sharp boundary between one individual and another, and our unremarkable, unchallenged notion of the individual loses meaning, as individuality itself becomes a matter of degree. No polyp is an island, entire of itself; every polyp is a piece of the whole, a part of the colony.

The gorgonian was a striking prize, but the dredge had not so far been uniformly successful. It was still plagued by its habit of filling too easily with mud, and it seldom captured any large agile animals—no deep-sea fish had yet been caught. At this point, someone suggested that any ordinary beam trawl, as commonly used in the deep-sea fishery, might be more effective. Most of the printed sources are coy about the source of this suggestion, but Murray states explicitly that it was made by Captain Nares: the naval officer giving technical advice about dredging to the naturalists! Even worse, it worked. The trawl was deployed that afternoon and came up full of animals—including the first deep-sea fish, eyes popping from the release of pressure. From this station on, both trawl and dredge were used, the trawl becoming increasingly the instrument of choice toward the end of the voyage.

Neither dredge nor trawl could be closed, so either might catch animals (quite efficiently) as it descended or (less efficiently, especially if half-filled with mud) as it was recovered. It was often difficult to judge whether an animal had really been caught on the bottom, unless it were something like a sea urchin or a clam that could live nowhere else, and in fact we now know that many of the animals recorded by the naturalists must have been captured far from the bottom, and even close to the surface. Even worse, any animal that had been captured on the bottom stood a good chance of being washed out of the net as it was being pulled up through several miles of water. The dredge and trawl used on the *Challenger* were clumsy devices that were bound to give an imperfect and misleading impression of what actually lived on the sea floor. An improved device called the epibenthic sled skims over the surface of the ooze, capturing many more animals, and can be closed at the end of its run so that nothing is washed out later. Remotely operated vehicles with cameras have been used more recently to give a direct and vivid picture of the sea-floor community and show many features, such as tracks and burrows, that no dredge could reveal. But the epibenthic sled was not invented until the 1960s (by Robert Hessler and Howard Sanders of the Wood's Hole Oceanographic Institute), and images alone, however vivid, cannot substitute for specimens in the hand. The steam-age technology of the *Challenger*, crude though it was, provided the foundation on which our knowledge of the deep sea was to be built for the next 100 years.

Station VII: 31 January 1873

35°20′ N, 13°4′ W; off Madeira, 2,125 fathoms

Challenger came alongside the jetty beneath the huge bastion of Gibraltar on 18 January for coaling. This was a filthy job, as always, the men working with handkerchiefs over their mouths in a vain effort to prevent the dust from caking mouth and nostrils, raking the coal in the bunkers to lay it flat, and checking that it was dry; damp coal can combust spontaneously, the ultimate threat to a wooden vessel. The officers and scientists, of course, went for a run ashore to visit the apes and the guns.

Pennatulids

A week later she left, steering west and south for Madeira, above a sea floor that slopes gradually at first then plunges down to the deep plain off Casablanca on the African coast. This was an easy time, loafing along, sounding, and trawling for seven or eight hours every day. The trawl was now beginning to bring up a regular procession of animals: glass sponges, polychete worms, a variety of echinoderms, and a few fishes. By 31 January there was more than 2,000 fathoms beneath the hull, and the trawl skimmed

Fig. 1.

Figure 6 *Pennatula* from Station VII

Source: A. Von Kolliker (1880), Report on Pennatulida. Zoology, Part 2, Plate 1. Bound in Reports, vol. 1.

over the blue mud of the deep seabed. When it was swung aboard that day it disgorged stalked animals some three feet long, looking something like gigantic phosphorescent asparagus stalks, but crowned with a tuft of tentacles. These were sea pens or pennatulids, another group of soft corals, related to the gorgonians and organized in the same way, as a colony of polyps making up a compound animal. Their name is an anachronism. Two or three hundred years ago, pens were made from feathers, the quill being cut, slit to make a nib, and dipped in ink. Pennatulids look a little like a quill pen: the animal consists of a long stalk, anchored into the sediment at one end by a bulb or holdfast and terminating at the other end in a feathery tuft. They look even more like individuals than gorgonians because their polyps are differentiated rather than all being the same. The primary polyp that founded the colony forms the stalk; some of the progeny that bud from it may be capable of feeding and reproduction, but others are specialized to ventilate the whole organism by creating currents of water without ever forming gonads. In this way, a colony of polyps may differentiate as an integrated whole by the division of labor among different kinds of polyp, a simple way of building a large animal from small units. This sea pen was a new species and was eventually named in honor of Wyville Thomson, *Umbellula thomsoni*.

The end of the beginning

Challenger sailed up a slope of volcanic mud and spent the first half of February at anchor or cruising around Madeira and Tenerife. The ship coaled; the men came back from shore leave staggering drunk from the strong Madeira wine; the naturalists pottered around exploring the islands or occasionally dredging in shallow water close to the shore. As they pottered, the scientists reflected on the voyage so far. It had not been unsuccessful, certainly, but it had been rather uneven. The apparatus as a whole had worked, but it had not worked very well, and on four occasions either sounding gear or dredge had been lost, together with thousands of fathoms of line. The trawl had been an unexpected and still unfamiliar addition to the program. All in all, it was agreed to regard the voyage so far as a prologue to the cruise proper, a trying-out period in which the new machinery was tested and the crew became accustomed to operating it. Having decided this, everyone (that is, the scientists and commissioned officers) enjoyed a ball hosted in their honor by the British Consul at Santa Cruz (perhaps they too came back staggering drunk, although the books do not say so) and prepared to set off across the Atlantic on the following day.

Challenger steamed out of Santa Cruz Bay on St. Valentine's Day and once clear of land set sail westward across the Atlantic toward the West Indies. The previous system of numbering the stations was dropped, and the next station would be designated number 1. It was the end of the beginning.

6

First Leg

The First North Atlantic Transect, February–March 1873

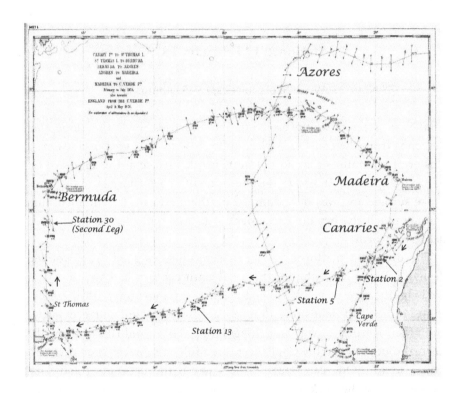

Challenger sailed south toward the flat Canary Basin to catch the trade winds that would blow her west to North America. The wind was fair and the weather kind, but the passage was slow because of the daily sounding and dredging. The ship had to be stopped for this, and stopping the ship meant furling the sails, and furling the sails meant hands aloft to roll up the sails and tie them down, and this might seem a light labor unless you had to climb 100 feet above the deck, edge out along a spar with only a footrope for support, roll up the huge sheet of canvas, and tie it to the

Full Fathom 5000. Graham Bell, Oxford University Press. © Graham Bell 2022.
DOI: 10.1093/oso/9780197541579.003.0007

spar. In calm water with the sun shining, furling the thin, patched, fair-weather sails was difficult enough; at night, with a sea running and the ship pitching and rolling, it was an almost superhuman task for the crew, clutching at the thick canvas, stiff with spray and ice, with numbed fingers. The *Challenger* could roll 30 degrees or even 35 degrees, so that a seaman at the end of a spar, if his hand should slacken, would fall directly into the sea, beyond any hope of recovery, and in a gale the ship would do this five times a minute. All this would come later. Meanwhile the ship sailed peacefully westward.

Station 2: 17 February 1873

25°52′ N, 19°22′ W; south of Canaries, 1,945 fathoms

Challenger steered southwest from the Canaries, over the slope that descends from the coast of Africa, quite close to (but unconscious of) a cluster of seamounts that do not quite break surface at the southern tip of the archipelago. At 3:30 in the afternoon of 17 February 1873 the trawl broke surface and was duly dumped on the platform. It contained mud and not much else beyond a fragment of a squid tentacle and a mutilated debatable worm.

The dredging platform had been carefully designed. The wooden deck of a Royal Navy ship of the nineteenth century had an almost sacred significance: it was scrubbed clean every morning with the holystone, a block of sandstone dragged around to grind out every spot and stain, washed down, and flogged dry. This would have been the normal routine when the *Challenger* was a fully gunned ship on the West Indies station. One can imagine the scene if the naturalists had been allowed to use the deck for their operations: they would have dumped a trawlful of mud on the spotless deck from time to time and rummaged around in their shirtsleeves, as though a stranger had shed a wheelbarrow from the garden onto the floor of your living room and then picked through it for interesting scraps. The sailors would murmur, of course, and the scientific gents would not notice or would ignore them, of course. This is why the unique and otherwise inexplicable dredging platform had been built, to save the deck and to preserve the harmony of the ship. It seems to have worked well enough to separate the science from the seamanship.

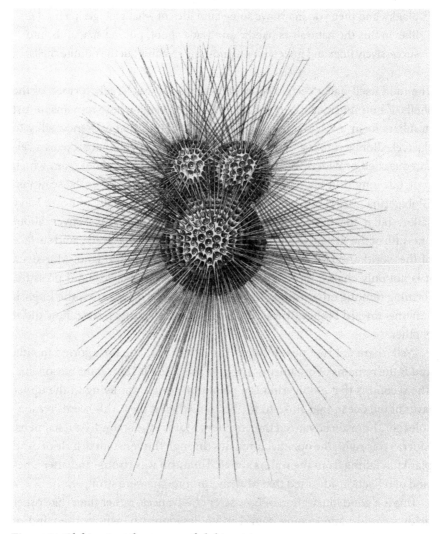

Figure 7 *Globigerina*, the source of globigerina ooze
Source: H. B. Brady (1884), Report on Foraminifera. Zoology, Part 22, Plate 77. Bound in
Reports, vol. 7.

Globigerina

Even when kept off the deck, mud did not much appeal to either crew or
officers. Here is Lieutenant George Campbell.

> The mud! Ye gods, imagine a cart full of whitish mud, filled with the
> minutest shells, poured all wet and sticky and slimy on to some clean

planks, and then you may have some faint idea of what globigerina mud is like. In this the naturalists paddle and wade about, putting spadefuls into successively finer and finer sieves till nothing remains but the minute shells.

The mud itself had a certain interest, however. It was largely formed of the shells of a minute marine organism called globigerina whose remains in vast numbers form a viscid, chalky ooze, almost liquid in consistency, silky to the feel. Globigerina is a foraminiferan, a kind of amoeba that secretes a calcareous shell provided with numerous pores (hence the name) from which protrude long thin filaments that catch bacteria suspended in the seawater. Globigerina ooze is not a very attractive substance, especially to Royal Navy sailors, but it does have a certain importance. In the first place, it covers about one-fifth of the surface of the earth. This is almost as much as the land surface of the world and twice as much as the area of agricultural land. Moreover, it is not only chalky, it *is* chalk, once compressed under heat and pressure, forming (among other things) the White Cliffs that stare across the English Channel toward France. It was also a first-class scientific puzzle: how did it get there?

Well, there are two possibilities. The first is that the ooze forms in situ, and is the remains of organisms living on the seabed. This seems reasonable. The second is that globigerina is a planktonic organism living in the upper layer of the ocean that sinks to the bottom when it dies. This seems reasonable too. Thomson advanced the first view, Murray the second. As it happens, Murray was right: the ooze accumulates through the constant drizzle of dead plankton falling from the waters above. Thomson was wrong, and after a period of reflection admitted that Murray's arguments were stronger.

This is a good illustration of how science advances, rather than (like other fields I could name) going round in endless loops blown by the wind of opinion. Thomson had a rational explanation in which he strongly believed, was countered by Murray's arguments, found them persuasive, reconsidered, and retracted his original view. Here is Thomson:

Since the time of our departure, Mr. Murray has been paying the closest attention to the question of the origin of this calcareous formation, which is of so great interest and importance on account of its anomalous character and its enormous extension. Very early in the voyage, he formed the opinion that all the organisms entering into its composition at the bottom are dead, and that all of them live abundantly at the surface and at intermediate depths, over the globigerina-ooze area, the ooze being formed by the subsiding of these shells to the bottom after death. . . . I had formed and expressed a very strong opinion on the matter. It seemed to me that the evidence was conclusive that the foraminifera which formed the globigerina-ooze lived on the bottom, and that the occurrence of individuals on the surface was accidental and exceptional; but after going into the thing carefully, and considering the mass of evidence which has been accumulated by Mr. Murray, I now admit that I was in error.

The globigerina controversy was a cameo of how science works.

Station 5: 21 February 1873

24°20′ N, 24°28′ W; Canary Basin, 2,740 fathoms

Challenger sailed west in fine weather, moving downslope into the deep water of the Canary Basin. Dredging on hard ground on 18 February brought up little more than a few stones and the stumps of an alcyonarian coral, a relative of the sea whip captured a month previously. Oddly enough, although the broken section of the coral was white, its surface was coated with some jet-black substance. Even more oddly, Buchanan found that the substance was almost pure manganese oxide. How it was formed was a puzzle to which Buchanan and the naturalists would return at intervals for the rest of the voyage. The dredge was lowered again on 19 February, but the line broke at depth, with the loss of a couple thousand fathoms of rope, plus the dredge, of course. On 21 February another attempt was made, using steam as usual to keep the ship head to wind, and this time the dredge was brought up successfully from over 2,500 fathoms. A new load of mud cascaded onto the deck, bearing very few animals indeed; a few small clams and not much else. The animals were not of much interest, but the mud was. It consisted of a red-brown clay, very different from the ooze found four days ago.

Red clay

It was the first sign of a pattern that was to be repeated throughout the voyage: globigerina ooze in deep water down to about 2,000 fathoms, red clay in deeper water still, down to abyssal depths. The clay covers about the same proportion of the ocean floor as the ooze; between them they cover nearly half the entire surface of the earth. What causes this worldwide transition from ooze to clay on the sea floor? Chemistry.

The shells and skeletons of many marine animals, such as corals, snails, crabs and foraminiferans, are reinforced by calcium carbonate, which the animals extract from seawater. They use specialized tissues to precipitate solid calcium carbonate from dissolved calcium and bicarbonate. When the animals die the process is reversed and their skeletons gradually dissolve. At the surface of the ocean this process never goes to completion, and most of the skeleton remains as a solid, such as a seashell or a coral reef. The balance between calcium carbonate (as a solid) and dissolved calcium is affected by pressure: calcium carbonate is more soluble at high pressure. Over most of the ocean floor, the rain of dead shells from plankton near the surface more than compensates for their dissolution, so that permanent deposits of ooze are formed. In very deep water, however, shells dissolve so readily that no ooze can form and the seabed consists of clay from which almost all the calcium has been removed. The clay itself is formed mostly of wind-blown dust, slowly sifting down through miles of water, entombing in the course of ages more interesting objects such as the teeth of sharks and the ear bones of whales, which the nets of the *Challenger* would bring to the surface later in the voyage.

This process is a little more complicated in practice because calcium carbonate exits in two crystal forms, calcite and aragonite. Some animals (such as foraminiferans) use calcite, and others (such as mollusks) use aragonite. Aragonite is slightly more soluble than calcite in seawater at given pressure. Consequently, ooze in relatively shallow water consists mainly of the shells of pelagic mollusks, whereas this is replaced in deeper water by ooze consisting mainly of shells of foraminiferans (the globigerina ooze). This is why there is a characteristic zonation of marine sediments with increasing depth, from mollusk ooze to globigerina ooze to red clay.

Wyville Thomson called this one correctly.

We conclude, therefore, that the "red clay" is not an additional substance introduced from without, and occupying certain depressed regions on account of some law regulating its deposition; but that it is produced by the removal, by some means or other, over these areas, of the carbonate of lime which forms probably about 98 per cent. of the material of the "globigerina-ooze."

The *Challenger* was beginning to produce a new map of the world, based on mud scraped from the bottom of the sea. Indeed, it was beginning to produce a new theory of the world.

Station 13: 4 March 1873

21°38′ N, 44°39′ W; the Dolphin Rise, 1,900 fathoms

As February turned to March the *Challenger* sailed a trifle south of west into the heart of the ocean, bowling along with the trade winds at six or seven knots, heading for the West Indies. Leaving the flatlands of the Canary Basin, she sailed over a vast field of hills, or low mountains, a few thousand feet in height, separated by pits and canyons. On 26 February the sounding lead found the bed of one such canyon, an unnamed gulch that runs for about 400 nautical miles westward from the rim of the Cape Verde Basin, at 3,150 fathoms, the deepest sounding so far. The dredge was lowered at the same time and brought up a scoop of fine red clay that, when stirred in a glass, remained suspended for days, showing how still the deep water must be for this impalpable dust to be compacted into clay. There was nothing else, and indeed little to show for the last week of dredging. Wyville Thomson writes,

> In the last few Stations the mud has been gradually altering its character, becoming less calcareous and less rich in Foraminifera. This mud consists almost entirely of reddish clay in a state of excessively fine division; it scarcely effervesces with acid, and there is an almost total absence of calcareous shells. No living thing could be detected except one or two Foraminifera with tests of fine brown grains.

The Mid-Atlantic Ridge

This was not very encouraging; the *Challenger* was traveling over a red clay desert apparently almost devoid of life. At the beginning of March, however, the soundings began to change. It would be natural to expect the floor of the ocean to be more or less level, on average, so far from land, or perhaps to deepen a little further toward the halfway point between North America and Africa. Instead, it began to rise quite abruptly, until soundings taken on 4 and 5 March showed that the sea was less than 2,000 fathoms deep. As the water became shallower the sediment changed too, so that for two or three days in early March the *Challenger* was picking its way across a plateau or ridge in the middle of the ocean floored by globigerina ooze. By 6 March the sea floor was falling away, and by 10 March the depth had dropped below 2,500 fathoms and there was red clay in the dredge again.

What the *Challenger* had found on 4–5 March 1873 was the Mid-Atlantic Ridge, the wholly unexpected rise of the sea floor in the middle of the ocean. We know now that the Ridge is the seam between tectonic plates, the American plates and the Eurasian and African plates, produced by basalt lava rising through convection from deeper in the mantle, pushing the plates apart and widening the Atlantic by about an inch a year. None of this could be imagined at the time, of course, nor for almost a century afterward; the Ridge was simply a fact of geography, like the Alps or Australia. This was not the very first time it had been found: the USS *Dolphin*, commanded by Lieutenant Otway Berryman, had reported an elevation of the seabed in this vicinity 20 years previously, and Nares (rather graciously) acknowledged this by naming it the Dolphin Rise. It was the *Challenger*, however, that first mapped this submarine mountain chain in the North and South Atlantic and established the Mid-Atlantic Ridge as a major feature of earth topography.

A blind lobster

The ooze brought up from the top of the Ridge on 4 March contained a few animals, including a rare lobster-like animal that interested the naturalists for two reasons. In the first place, it belonged to a family that had flourished some 150–200 million years ago during the Mesozoic era, alongside ammonites and dinosaurs, and was widely regarded as a sort of living fossil that still retained primitive characteristics of ancient kinds of crustacean. Second, although its ancestors in the Mesozoic were sighted, it was blind. It was completely blind, with no eyes and only rudimentary structures where its eyestalks should have been. It is tempting to interpret the loss of sight as an adaptation to life in the lightless depths of the sea, much as webbed feet aid paddling or colorful flowers aid pollination. This is not quite incorrect—it is

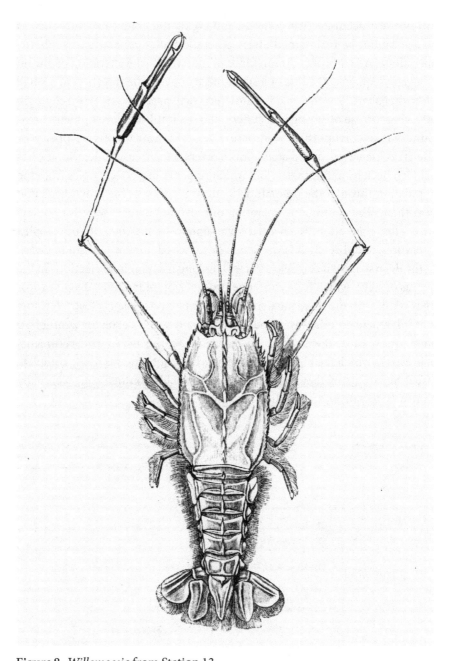

Figure 8 *Willemoesia* from Station 13

Source: C. Spence Bate (1885), Report on Crustacea Macrura. Zoology, Part 34, Plate 18. Bound in Reports, vol. 12. Also in Narrative, vol. 1, Part 2 Figure 181.

just not very accurate. Any complex and highly integrated structure such as an eye cannot be maintained intact generation after generation by inheritance alone. All sorts of common kinds of mutation—the alteration of one bit of a gene sequence, the loss of a small piece of gene, the insertion of a small piece that alters how the gene is translated—will produce imperfect copies in offspring, whose offspring will suffer likewise, until the developmental machinery is so corrupt that the structure can no longer function properly, or else is no longer produced at all. This irreversible process of deterioration is held in check by natural selection, which removes imperfect copies from the population through the premature death or sterility of the individuals that bear them. The continuous operation of purifying natural selection, generation after generation, is essential for maintaining the integrity of complex structures, and directly it is weakened or removed, they will begin to decay. In the deep sea, purifying selection will no longer act on imperfections in the organs of sight, which have consequently deteriorated in most deep-sea animals until, as in the blind lobsters, they have been completely lost. We know that it happens in this way, rather than sighted species evolving from blind ancestors, because the direction of evolution is given away by the rudiments of eyestalks which the blind species bear as the last remnants of the complete visual apparatus of their ancestors. This particular animal was eventually named *Willemoesia*, after Willemoes-Suhm.

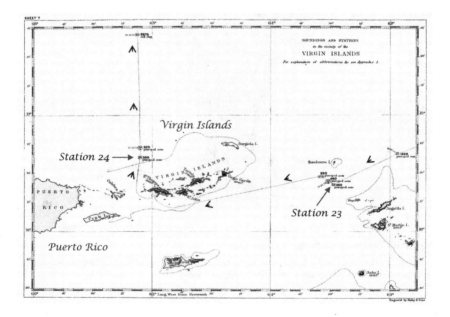

Station 23: 15 March 1873

18°24′ N, 63°28′ W; off St Bartholomew, 450 fathoms

The weather was by now fine and warm as the *Challenger* moved gently westward in fits and starts, dredging and sailing. Every 200 miles steam would be raised to maneuver the ship for dredging; then the screw would be raised and locked, and the fires banked; the sails were then unfurled and the ship sailed a further 200 miles; repeat indefinitely, or so it must have seemed. The off-duty officers would shoot at bottles or sharks, and smoke on the upper deck in the evening, now that the weather was warmer. The naturalists, getting into the spirit of shipboard life, donned blue serge jackets and white trousers for work on the platform. I imagine, although no records were kept, that Wyville Thomson, Murray, and Willemoes-Suhm would usually have formed the platform party; Moseley was uninterested in dredging and would be more likely to be sorting the animals collected on land, while Buchanan would be in the chemistry lab, where he was about to make a surprising discovery. The dredging operation itself was usually supervised by Captain Nares, although Tizard was responsible for the soundings. There was the occasional break in the routine. A Spanish brigantine with girls on board passed within hailing distance and was duly hailed (giving her longitude was the excuse). Flying fish fluttered and iridescent dolphins (the fish, not the mammal) glided through the sea. This was more like it.

Black pebbles

The sounding tube brought up red mud on 7 March, to the dismay of Wyville Thomson, who was learning that this red mud was usually devoid of life. Nevertheless, the dredge was put over, and sure enough contained little more than red mud when it was recovered and emptied onto the dredging platform eight hours later: a single bivalve shell and some odds and ends, shark teeth, and a few small black pebbles. It was arguably the richest single haul ever taken by a dredge.

The sharks' teeth were fossils, of course, coming from some long-dead animal whose corpse had plummeted down from the upper waters, plashing into the ooze. At first glance the pebbles were probably fossils of some sort too, perhaps the phosphatized remains of some sea creature. They were not, though: when Buchanan tested them in his chemistry lab, they turned out to be almost pure manganese. This presented a first-class academic riddle: how were they formed? It became an even more interesting academic question when, later in her voyage, the *Challenger* discovered extensive fields of these nodules strewn over large areas of the ocean floor. The riddle was soon solved: they were formed by manganese precipitated from seawater around a core that is often, in fact, a fossil, such as a shark's tooth or a fragment of a sea urchin. Not just manganese, either; the nodules contain nickel, cobalt, and copper too. At this point the discovery was no longer of purely academic interest. What the *Challenger* had discovered was not just a few curious pebbles; it had discovered the largest metal mine on Earth. More than a century later, we are still trying to work out how to exploit it. The effort will no doubt be worthwhile: there are some hundreds of billions of tons of metal locked up in the kind of nodules that Buchanan tested in his swaying onboard laboratory on *Challenger*, where he first saw the pale yellow-green flame of manganese.

The isolation of the dredging platform, above the main deck, was eminently practical; but it was also to some extent symbolic. It certainly emphasized the separation of the scientific gents from the crew. The crew were not entirely content with this. They had heard rumors, and had read accounts in the press, of how important the voyage was for science, which was surely worthy of respect, but they had not actually been directly told anything whatsoever. Perhaps these murmurs reached the bridge; at any rate, on 10 March Wyville Thomson addressed the crew to explain what they were about. He said that we know little of the sea floor that covers most of the surface of the Earth—previously thought that nothing could live in the deep sea, but we now know this to be false—importance for laying undersea cables—a British venture, not to be left to foreigners—unique characteristics of the deep-sea fauna. He closed by promising to address the crew again on progress made during the voyage. It was an important speech, reading in the scientifics, so to speak, to the crew. One can see in the mind's eye the bluff, tubby figure of Wyville Thomson explaining to the assembled seamen why they spent most of every fine day bringing in dredges full of mud from the ocean floor. The speech is not mentioned, oddly enough, in any of the contemporary accounts of the voyage, and I am relying on a single letter from Matkin. It may even have been given the previous week. At all events the promise to repeat the lecture was not kept, so there was no repeat performance, at least as far as the records show. The separation of scientists from crew was difficult to overcome or even to acknowledge; although Matkin calls the lecture "a great success," perhaps neither side really wanted a repeat.

A deep worm

The *Challenger* continued westward over a hummocky landscape of red clay in deep water that was almost, but not quite, devoid of animals. On 12 March the dredge was put over, and at a depth of almost 3,000 fathoms passed through a grove of tubeworms, *Myriochele*. In life, these would have been embedded in the sediment, protected by the tough flexible tubes they make, but projecting above the surface to filter out the minute organisms that live near to the bottom in deep water; they are about as long as a middle finger, but thin as a pencil lead. The dredge came up full of red mud, but plastered with tubes, some with the animal still inside. The worms themselves had been found before, and were related to similar species familiar from coastal waters. Wyville Thomson makes much of this haul, because it showed conclusively that animals inhabit the deepest water that had yet been trawled. The animals were undoubtedly from the sea floor, embedded in clay, and the last vestige of the idea that animals could not live at great depth had been dispelled. Thomson deserved his triumph, and was perhaps relieved that he had gained his point and justified (at least from the point of view of a zoologist) this long and expensive cruise. Yet some doubts might still linger in a more skeptical, or less committed, mind. If animal life were not completely extinct at depth, it was certainly thinning out. Was it possible that there might after all exist a zero of animal life, but at much greater depth than had been anticipated? This was a question that could not be answered until *Challenger* searched the ultimate depths, which lay a year in the future and half a world away.

Challenger passed over the Puerto Rico Trench, which runs like a ditch across the northern edge of the Lesser Antilles, on the following day, obtaining a sounding of 3,025 fathoms. Despite the depth, the ship was now approaching land, with a complex system of valleys extending out into the ocean from the high plateaus where the islands at last poke through into air. On 14 March the water began to shoal rapidly as *Challenger* approached the Lesser Antilles, and at daybreak on 15 March St. Bartholomew was visible ahead. In the space of these two days the sea floor had also changed completely, from red clay to a thick ooze consisting mainly of the shells of small snails.

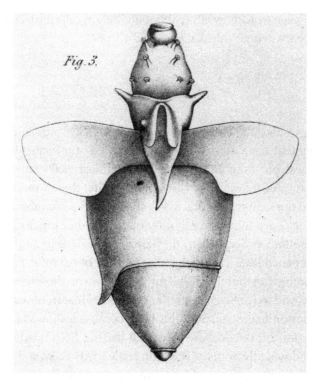

Figure 9 *Dexiobranchaea* from Station 23
Source:, p. Pelseneer (1887), Report on Pteropoda Gymnosomata. Zoology, Part 58, Plate 1. Bound in Reports, vol. 19.

Snails with wings

Snail shells are not the most obvious objects to find in marine sediments. Snails abound on hard surfaces in coastal waters because they are scraping animals. They use their muscular foot to crawl over rocks and seaweed, rasping away the slimy film of bacteria and algae that coats the surface with a very hard tongue-like organ, the radula, and swallowing the fragments. This is not a method that works in soft sediments, so most of the mollusks that *Challenger* dredged from depth were filter-feeding bivalves. There was no mystery where the shells came from, however: they came from an extraordinary group of pelagic mollusks, the pteropods, which swarm in the upper waters of the ocean throughout the world. Like the snails that live near the coast or on land, they have a foot, but instead of using it to creep slowly along they use it to fly. Their foot has evolved into a pair of thin winglike structures that can be rapidly fluttered up and down to propel the animal through the water like a tiny butterfly. Wyville Thomson waxes (it is the only word) quite poetic about them: "Multitudes of the little things may now and then be seen on the surface of the water, fluttering their wings and glittering in the sunshine, to be compared with nothing more aptly than a congregation of the more dressy of the bombyx moths, as one sometimes comes upon them on a sunny morning, just after a family of them have escaped from their chrysalises." Their body is encased in a thin delicate shell, formed as a simple cone or coiled into a spiral. When its brief life is over, the dead animal drifts very slowly downward—they are almost neutrally buoyant—while its flesh is consumed by fungi and bacteria, until only the shell remains. It is this continual thin rain of tiny dead shells that forms the pteropod ooze over which *Challenger* was sailing. There is not much of it, however, despite the cosmopolitan distribution and enormous abundance of the living animals: only about 2 percent of the Atlantic, and perhaps one-tenth as much of the Pacific, is floored by pteropod ooze, all in shallow water, usually near land. The reason is that their shells are made of the aragonite form of calcium carbonate, which dissolves more readily in seawater than the calcite used by forams, especially at depth. The result is that the pteropod ooze is replaced below a certain depth—the aragonite compensation depth, at which dissolution just balances accumulation—by globigerina ooze, which is in turn replaced, at the calcite compensation depth, by red clay.

Pteropods are blind, like many deep-sea animals. But they do not live in the deep sea—on the contrary, most inhabit the upper few feet of the ocean. They are blind, nevertheless, for the same reason that deep-sea animals are blind, because their way of life does not require and would not be facilitated by sight. They feed by secreting a large web or net of mucus and consuming the algae and bacteria that it traps. There is another group of pelagic mollusks, the heteropods, which have independently evolved a winglike modification of the foot, and which also contribute to the formation of calcareous oozes in shallow water. Heteropods have a quite different way of life: they are active predators that catch small crustaceans and fish larvae; a blind heteropod would be unable to feed, and consequently the group has evolved complex image-forming eyes equipped with a large optic nerve. A few other kinds of snail live in the upper waters of the ocean; for example, *Janthina* simply floats along, buoyed up by a raft of air bubbles enclosed in a thin sheet of chitin. It has a simple way of life: from time to time it bumps into a jellyfish, such as the by-the-wind sailor *Velella*, and eats it. Wind and current keep *Janthina* and its prey more or less together, and lacking any means of active locomotion sight would be pointless. *Janthina* is blind.

Sharks

The sharks that followed the ship were not blind, and snapped up anything edible that fell or was tossed overboard. Their numbers increased as the ship closed the land, and the daily halts for dredging gave the crew a good opportunity to catch them, using the shark hook issued to all naval ships—basically, a butcher's hook attached to a length of chain. Baited with a chunk of salt pork and floated off with a barrel a few yards from the ship's side, it would soon take a small shark of five or six feet long. There was no nonsense about playing the fish; it was hauled alongside, a running bowline slipped over its snout to the root of the tail, and then hoisted inboard by the derrick to twist above the deck. Sailors hate sharks, with good reason if you imagine falling overboard while they are about, and give them a hard cruel death with axe and knife. One was disemboweled, alive, and being pregnant (most large sharks are livebearers, and some even have a placenta) dumped half-a-dozen unborn young onto the deck, which made an agreeable addition to breakfast. Fetal shark might not be your idea of starting the day, but if your usual meal were hardtack with most of the weevils tapped out first, it might begin to appeal. They did not catch the big one, though, a 20-foot monster that had shadowed them for a thousand miles and grabbed anything thrown overboard "like a lawyer" (Matkin's phrase). It would sniff at the salt pork but turn away without taking the bait, which was doubtless why it had survived to be 20 feet long.

Station 24: 25 March 1873

18°38′ N, 65°5′ W; Virgin Passage, 390 fathoms

Challenger was delayed by fog around Sombrero, a small uninhabited island that marks the point where the island chain turns to the west, before docking in the harbor at St Thomas, in what is now the British Virgin Islands, on 16 March. She stayed here for the next week or so, taking the opportunity for a run on shore and grubbing around in the shallows. Herbert Swire was captivated by the islanders, especially the women; Lord George Campbell found the place ugly and disagreeable, especially the women; Mosely attended to the animals and plants without noticing the people much, although St Thomas did not have a lot to interest the naturalist, having been almost completely cleared for sugar cane, back in the days when slavery made the crop profitable. Toward the end of their stay they were told of a ship in distress, the *Varuna*, which had been dismasted in a storm after sailing from New York and abandoned by her crew. They found the ship, but also found it occupied by a prize crew from another ship who had rigged jury masts and sailed the derelict to within 15 miles of St Thomas. The situation was a little delicate. The prize crew were in actual possession and had made the ship navigable, after a fashion, and so they had the rights of salvage. They might lose these rights if they were towed into harbor by *Challenger* and were understandably reluctant to accept any assistance. Nares eventually persuaded them that *Challenger* had no intention of pressing a salvage claim, and towed *Varuna* back to safety at St Thomas. *Challenger* set sail for Bermuda the following day.

Accidents

A sailing ship is a very dangerous workplace. A boy had already fallen from the mainmast, fortunately into the sea, where he was rescued, brought aboard, and quickly recovered. On another occasion, a heavy sea struck the rudder, flinging two men against the wheel when the wheel-ropes broke. The same day, a seaman fell from the main trysail mast to the deck and was badly hurt. The most serious incident occurred on 25 March, when the dredge fouled in shallow water. When this happens, the dredge rope is drawn taut by the donkey engines and by the inevitable movement of a thousand tons of ship. The tension on the dredge rope puts most strain on the block at the yardarm and on the first leading block, attached by a span of rope to the upper deck. The accounts differ in detail about what happened next. All are agreed that the attachment of the leading block failed, either because the rope snapped or because an iron hook broke. The heavy block whipped through the air, hitting a boy named William Stokes on the side of the head and throwing him to the deck. He was carried below but died four hours later without regaining consciousness. I imagine that any modern workplace would have been immediately shut down, all operations stopped, the authorities brought in,

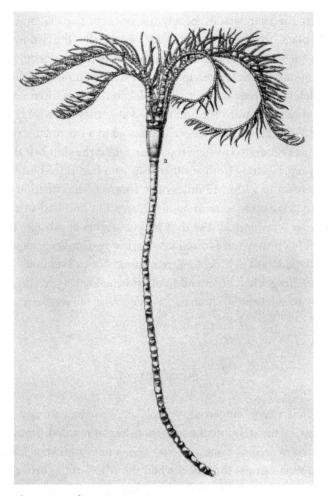

Figure 10 *Rhizocrinus* from Station 24

Source:, p. H. Carpenter (1884), Report on Crinoidea. First, Part. Stalked Crinoids. Zoology, Part 32, Plate 9. Bound in Reports, vol. 11. Also in Narrative, vol. 1, Part 1 Figure 119.

everything left as it was so that the cause of the accident could be ascertained, and of course any necessary changes made to ensure that nothing like this could ever happen again. The Navy, however, was used to taking casualties. The block was reattached, the trawl brought in, and the catch deposited as usual on the dredging platform for the naturalists to sort through. They found quite a lot of animals. The ship then steamed on, and later in the day, with William Stokes dying below of a cerebral hemorrhage, the trawl was put over again in 625 fathoms.

A stalked crinoid

There were just two animals in this trawl, both stalked crinoids called *Rhizocrinus lofotensis*. Crinoids are echinoderms related to starfish and sea urchins and share their fivefold symmetry, although each of their five axes may divide several times to produce dozens of arms, and each arm bears a fringe of long pinnules, making it look something like an ostrich feather. They feed by trapping minute organisms in a sticky mucus (like pteropods, although entirely different structures are involved) secreted by the pinnules. Most living crinoids are feather stars, which often live on hard surfaces in shallow water with strong currents, such as the sides of gullies in coral reefs, where they can trap the plankton as it flows through their pinnules. On soft sediments this would not work very well, because the pinnules would be clogged by mud, so crinoids living on mud or ooze usually have a long stalk, consisting of a stack of five-sided calcareous ossicles, which raises their body well clear of the bottom. Stalked crinoids are very common as fossils, usually as disarticulated ossicles strewn in the rock, from past ages when crinoids were much more abundant than they are nowadays. They fade from the fossil record during the last days of the dinosaurs, and until G. O. Sars dredged a stalked crinoid—the same species found by the *Challenger*—from 360 fathoms off the Lofoten Islands in 1864 they were thought to have become completely extinct. Indeed, this discovery nourished the immensely seductive idea that the deep sea might conceal many other supposedly extinct groups of animals. This was one of the reasons for the *Lightning* cruise—the principal motive for the cruise, according to Carpenter—which duly found *Rhizocrinus* in 530 fathoms north of the Outer Hebrides. The same species was also dredged from 2,435 fathoms by *Porcupine* west of the Bay of Biscay. Perhaps the deep sea really was the attic of the world, stacked with treasures from bygone ages.

The attic of the world?

The attic of the world—what power lies in an analogy! Basement might be even better, or the lumber room of the Earth—both have been used to express the idea. It is as though there were someone who regularly tidies up, collecting all the old, worn-out things and carefully storing them out of the way, in case they should ever be useful again, or just for sentimental reasons, like a faded photograph or a child's first shoes. We know full well there is not really anyone there, but the analogy nevertheless lassos the imagination. Besides, did not the blind lobster that *Challenger* had found earlier in the month belong to a group that was thought to be long extinct? In fact there was more evidence than the naturalists knew, in the shape of fossils resembling another of the animals from the dredge that would only be discovered a century later.

These fossils are found on the flat gray rocks of Mistaken Point, far to the north of St Thomas near the southern tip of Newfoundland, which look out like a tilted tabletop across the Atlantic toward Europe. It is an unexpected and slightly unsettling place. You may walk as you wish on most of the coast of Newfoundland, always supposing that you can reach it in safety on this storm-swept shore. Mistaken Point is different. You are forbidden to drive or cycle on these rocks, or even to walk across them without special permission. Above all, you may not attempt to collect any of the fossils that lie strewn over the slabs, because this is one of the very few sites in the world that bears the remains of the strange organisms that flourished just before the first recognizable animals appeared, some 565 million years ago. By a fluke of history they were buried in fine-grained volcanic ash that has preserved their bodies as highly detailed three-dimensional impressions, almost like a latex mold. Nobody is quite sure what they were: they might be the very early ancestors of modern animals or, perhaps, a separate and unrelated evolutionary experiment that died out without leaving any descendants. They vary in appearance, but one common type consists of a frond borne on a stalk that was anchored in the sediment by a disk-shaped holdfast. It looks a little like a fern frond pressed into the rock. Presumably it lived by filtering out minute organisms suspended in the sea. The detailed structure of the frond is unique, and not easily compared with any existing kind of animal. The gross anatomy, on the other hand, is quite familiar: it looks like the pennatulid that the dredge had brought up off Madeira two months previously.

There are plenty of reasons to be skeptical: 625 fathoms is no great depth, and any kind of animal living in ooze might need a stalk to keep it from choking. It might be best to defer judgment and wait to see what other strange animals the *Challenger* was to find in the depths of the sea.

7

Second Leg

The Sargasso Sea and Gulf Stream, March–May 1873

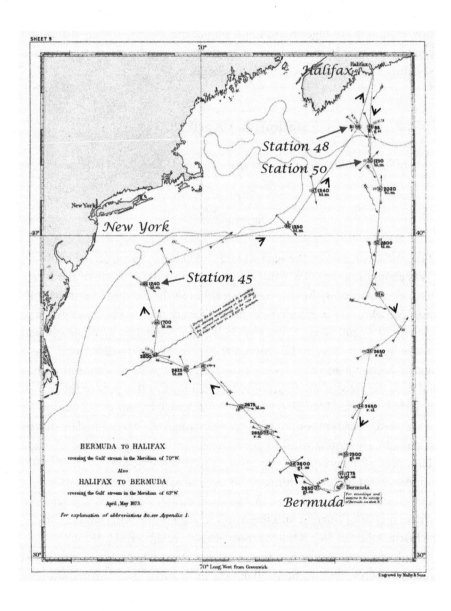

Full Fathom 5000. Graham Bell, Oxford University Press. © Graham Bell 2022.
DOI: 10.1093/oso/9780197541579.003.0008

William Stokes was buried at sea, in the traditional manner, sewn into his hammock with two round shot at his feet, on the following day, 26 March. As it happened, he was buried in the deepest water that *Challenger* had so far encountered, 3,875 fathoms in the Puerto Rico Trench. The *Challenger* continued north toward Bermuda across the flat and almost featureless Nares Plain, with depths close to 3,000 fathoms, floored by red clay. This gave way as she sailed north to a gentle hummocky slope gradually shallowing toward the Bermuda Rise, although still with soundings of more than 2,500 fathoms. Three casts of the dredge brought up a single animal, a red shrimp that might or might not have been caught on the bottom. It was the red clay desert that the naturalists were now beginning to recognize.

Station 30: 1 April 1873

29°05′ N, 66°01′ W; approaching the Bermuda Rise, 2,600 fathoms

Sargassum

At the surface, however, the sea was far from being a desert; on the contrary, it was abounding with life. Murray was out most days in one of the boats, often accompanied by Moseley, poking about in the floating islands of gulf-weed. Moseley was in his element here: "We are now at last amongst large patches of the gulf-weed. From the deck at noon patches, some nearly half an acre in extent, were to be seen on the surface in every direction." These patches housed a zoo of unusual animals, although none were more unusual than the weed itself, floating in large clumps in this part of the ocean, buoyed by air-filled bladders. Floating is not usually a successful way of life for seaweed, because sooner or later any floating raft of weed will be driven on shore by current or wind to decompose. It will only be successful when a large area of sea is boxed in by currents so that the weed is whirled round and round without ever coming on shore. The Sargasso Sea is boxed in by a clockwise succession of the North Atlantic Current, the Canary Current, the North Equatorial Current and the Antilles Current, forming a huge eddy in the North Atlantic. This is the Sargasso Sea, where the weed lives.

Think of a bush growing in the water that starts off as a little cutting and proceeds to grow from its tip in all directions, branching as it grows, with the tip of each branch growing in turn. This will soon form a ball of leaves. Now the older parts—the original stem of the cutting and its first branches—begin to senesce and die. They slough off and decay, liberating new, young growing tips, which drift away and repeat the process of growth and senescence. This is the life cycle of the gulfweed, *Sargassum*—not of all its species, indeed, most of which are perfectly normal coastal seaweeds attached to rocks with a holdfast, but of two species that have evolved into entirely pelagic plants that float like duckweed in a pond within the two million square miles of the great North Atlantic Gyre, northeast of the Antilles. Every plant attracts animals as a bush attracts birds.

The first to arrive is a hydroid called *Clytia*, another colonial organism whose polyps rapidly bud into a network that covers the growing plant, using it as a solid base in mid-ocean to catch their tiny prey. A slower but more solid bryozoan, *Membranipora*, then begins to extend its white calcareous colonies over the leaves of the seaweed, forming large white blotches. It may be joined by a tubeworm with a spiral white housing. As it ages, the bright yellow of the young weed becomes a darker brown, decorated with patches of the organisms growing on it. Other animals join them: flatworms, shrimps, and crabs. All resemble the weed, with yellow-brown bodies covered with white blotches, to make it difficult for their enemies, who are many, to find them. The enemy of all is the gulfweed angler, *Histrio*, a small and shabby fish covered with a camouflage of rags and tatters of skin that clambers among the weed in a most unfishlike manner, using modified pectoral fins, in search of its prey. It has a fishing lure, a little flap of skin at the tip of one of the spines of its dorsal fin, that it waggles when it finds some likely victim, such as a small shrimp. If this gets it close enough, it sucks in its prey—quite literally; its mouth expands to 10 times its normal size in a few milliseconds, creating a current of water that plucks the shrimp from its hiding place in the weed. The only enemy of the angler, at least on this occasion, is the dip net of naturalists like Murray and Moseley, who added dozens to their collections.

Now, if the surface of the Sargasso Sea is a fertile meadow of floating aquatic plants, why is the seabed below a desert? Surely one would expect the detritus of the surface of the sea to furnish a rich tilth for its depths, as the leaves of forest trees rot into soil. Yes, but a dead leaf falls several meters to the ground in a few seconds, and not much happens to it on the way; it can then proceed to decompose. A fragment of *Sargassum*, or a dead phytoplankton cell, or the fecal pellet of a crab or shrimp, equally fall downward, but they fall through a much denser medium for several kilometers before they reach the ground, and a great deal may happen on the way. The speed at which a particle sinks in seawater depends on its density and size; very roughly, about 40–50 meters per day for plant fragments or fecal pellets, scarcely 0.5 meters per day for phytoplankton. A small fragment of *Sargassum*, a millimeter or so across, sinking passively through the water, will take 50–100 days to reach the sea floor four kilometers down, and by that time there will not be much of it left; generations of bacteria and fungi will have consumed anything remotely edible. This is why the dredge was not bringing up a rich community of burrowing animals with the mud; there is simply not much to eat in the deep sediment.

Challenger skirted the edge of the Bermuda Rise, coming into shallow water floored by coral mud on 4 April. Bermuda itself was one of the least favorite landfalls for sailing ships, being surrounded by a tortuous maze of channels through the coral with every prospect of going aground at any moment and tearing the bottom out of the ship. *Challenger* was brought in safely by a pilot ("a Darkie" says Matkin) and anchored in Grassy Bay, where she joined the flagship *Royal Alfred* and the rest of the fleet in the naval yard for refitting and victualing. That night the schoolmaster, Ebbels, was heard breathing loudly after he had retired, and could not be roused; he died before the doctor could be found, and was buried the next day, only a few hours later, in the Bermuda cemetery, on his thirty-seventh birthday.

The Gulf Stream

The *Challenger* spent the next two weeks in Bermuda, with the usual excursions on shore and a little light dredging. She left on 21 April, passing over a very shallow bank, scarcely 30 fathoms deep, floored with pebbles studded with small branching corals. It was a seamount, the remains of a submerged volcano, well known to local fishermen but missing from the Admiralty charts, and was promptly named Challenger Bank. From here she turned west, sailing over a steep slope leading to much deeper water of more than 2,500 fathoms again. At first she made for New York, but a fresh gale drove her back, rolling badly. There was a brief respite on 28 April, a fine day when the dredge could be lowered, but came up with little more than ooze. During the night the breeze sprang up again, however, and became so strong that a tiller rope broke and the three helmsmen were thrown aside by the wildly spinning wheel, fortunately without serious injury. Dredging was out of the question, but an odd lump of whitish clay that turned up in the chemistry lab among soil specimens from Bermuda puzzled the scientists, who found it difficult to explain how it could have been formed. They were even more puzzled when it mysteriously disappeared, and they searched for it all over the ship. Whether or not they were glad when they found it is difficult to say: it turned out to be a ball of pipeclay, reclaimed by the marine who was using it (as was the usual practice at the time) to cover up the stains on his white uniform breeches.

The following day the temperature at the surface of the sea abruptly rose by six degrees Celsius as the ship crossed the southern edge of the Gulf Stream. This is perhaps the one ocean current that everyone has heard of, transporting warm water from the Caribbean across the Atlantic to northwest Europe, making it possible to grow crops in Britain, Ireland, and southern Scandinavia, and even (after a long chain of causation) to mount major oceanographic expeditions. One can only imagine the disappointment of the first sailors to cross the Atlantic when they found, at the same latitude as Dublin and Liverpool, only the ironbound coast of Labrador. It is quite a narrow ribbon of water that *Challenger*, running in the face of a fresh easterly breeze, crossed in a couple of days, marking its northern border as the color of the sea changed from a deep clear blue to cloudy green.

Station 45: 3 May 1873

38°34′ N, 72°10′ W; approaching Hudson Canyon, 1,240 fathoms

Sea serpents

Challenger was now skirting the land, sailing over a plain with a rather uniform depth of about 1,500 fathoms. The dredge brought up a bluish mud, globigerina ooze mixed with sediment from the land, and a good collection of starfish and sea urchins caught in the tangles. There was a strange creature in the mud, which, apart from a similar animal caught the previous day, the naturalists had not previously encountered on the voyage. Matkin records them as "2 small sea serpents." Campbell more soberly writes of "a worm the size of a small eel." This is unlikely to have been a young sea serpent, but it was probably not one of annelid worms found here either, since most were

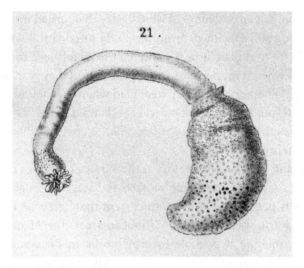

Figure 11 *Phascolion* from Station 45
Source: E. Selenka (1885), Report on Gephyrea. Zoology, Part 36, Plate 4. Bound in Reports, vol. 13.

in fragments and the specimens that came up whole would have had a prominent fringe of swimming legs. It was probably the "very long Sipunculus"
that is noted in the Station-book but seems to have been subsequently lost.
Sipunculids are not very familiar animals, except to zoologists, because, although they are not rare in shallow water close to shore, they are generally
small and lie concealed in holes and crevices during the day; besides, they
are not very attractive, looking too much like something dangling from a
baby's nose. They have a simple and effective body plan, however, essentially
consisting of a fluid-filled cavity enclosed by a muscular body wall. The muscles of the wall are arranged in two layers, one wrapped around (the circular
muscles) and one extending from end to end (the longitudinal muscles).
Contracting the circulars necessarily extends the longitudinals because the
fluid of the body cavity is incompressible, making the body long and thin;
likewise, contracting the longitudinals makes the body short and fat. In this
way the animal can change its shape by using its fluid-filled interior as a hydrostatic skeleton that allows the two muscle layers to oppose one another.
When a wave of muscle contraction passes along the body, the shortened base
of the body acts as an anchor against which the elongating head section can
be thrust forcefully through sediment, which is how worms of simple construction can burrow. Sipunculids use their hydrostatic skeleton for another
purpose. The interior of the worm is divided into two main compartments,

a short, thick-walled body compartment and an elongate, thin-walled proboscis. Contracting the circular muscles of the body compartment extends
the proboscis from the hiding place of the worm out into the open, where
it can grope around for food on the surface of sand or sediment. Any edible
particles it encounters, such as algal cells, are swept up by cilia, mixed with
mucus and channeled down a groove into the mouth. Many quite unrelated
kinds of animal—crinoids, for example—use a similar kind of mucus-ciliary
feeding device.

Challenger sailed westward, just off the edge of the Georges Bank, and,
unknowingly, across the New England Seamounts. These form a chain of
about 20 underwater hills that rise about 2,000 fathoms above the seabed.
They are extinct volcanoes formed by a hotspot that burns a hole through
the overlying crustal rocks. The oldest was formed about 100 million years
ago, and the chain was formed subsequently as the continent drifted slowly
westward. Each new volcano cooled and contracted as the hotspot moved
away, sinking into the sea, so that all are now at least 500 fathoms beneath the
surface. *Challenger* probably passed between Bear and Physalia Seamounts
on her way to Halifax, having by now given up the original plan of calling at
New York.

Station 48: 8 May 1873

43°04′ N, 64°05′ W; Le Have Bank, 51 fathoms

The crinoid plague

Challenger sailed over the cliff about 3,000 feet tall that separates the banks
from the continental shelf and anchored in 50 fathoms of water on the Le
Have Bank so that the crew—officers and men—could fish for cod. The
naturalists put the dredge over and found the bag empty (the seabed here is
bare rock) but the tangles thick with echinoderms, especially crinoids and
brittle stars. The crinoids were a stalkless species of *Antedon*, the feather
star, which lives on hard surfaces; more precisely, they have only a rudimentary stalk, equipped with prehensile hooked structures that can anchor it to
rock or weed. They can crawl quite rapidly and even swim short distances
by thrashing with their arms. These particular *Antedon* were heavily

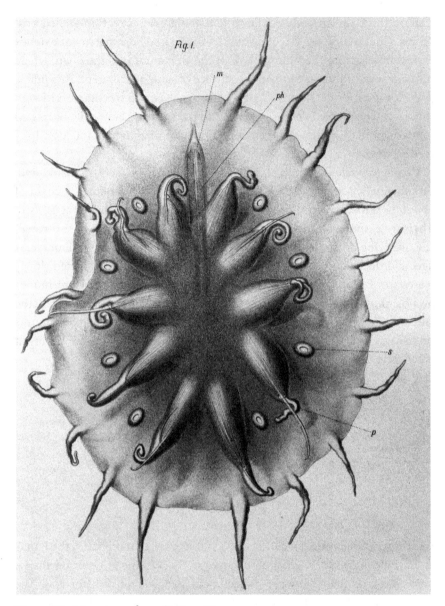

Figure 12 *Myzostoma* from Station 48

Source: L. Von Graff (1884), Report on Myzostomida. Zoology, Part 27, Plate 6. Bound in Reports, vol. 10.

infested with *Myzostoma*, the plague of crinoids. Mucus-ciliary feeding is a simple way of life, but an unprotected food groove is an open invitation to freeloaders like myzostomids. These are small oval animals with five pairs of short legs that crawl on the surface of crinoids and stick their pharynx into a food groove to suck out the contents. There is nothing the crinoid can do about this—it is in the position of a knight in armor attacked by a leech. Other kinds have become invasive and crawl into the gut or gonads for a more intimate meal. For the *Challenger* naturalists they were a zoological puzzle because they do not seem to fit into any of the main groups of animals; most authorities interpreted them as a highly modified kind of annelid, but others thought they might be allied to flatworms or crustaceans or even rotifers. The most recent molecular studies put them back into the annelids, but the historical level of uncertainty shows how difficult it is to understand animals that have evolved as parasites and become quite unrecognizable after long ages of modification and specialization.

Brittle stars

The fishing was successful. Nares bought bait from a nearby fishing boat, and in those days the banks were still floored with cod, an apparently inexhaustible resource. It was hard work pulling up the big fish from 30 or 40 fathoms down with handlines, but a change from the usual routine and a welcome change of diet too. The men fed on the cod for dinner, tea, and supper; the cod in turn had been feeding on *Ophiopholis*, one of the ophiuroids, the brittle stars, caught by the dredge. The stomachs of the fish were crammed with them, perhaps because there is not much meat on an ophiuroid. These are quite familiar animals, looking rather like a starfish except that their central disk is more emphatically separated from five slender arms. They will scuttle rapidly away when you turn over a rock in the intertidal, and if you pick one up they will still scuttle away, leaving you holding the arm they have abandoned. An ophiuroid is covered by a sort of mail coat of calcareous plates and disks, which may be colored black or red or gold, giving it more the appearance of a fantastical piece of jewelry than an animal.

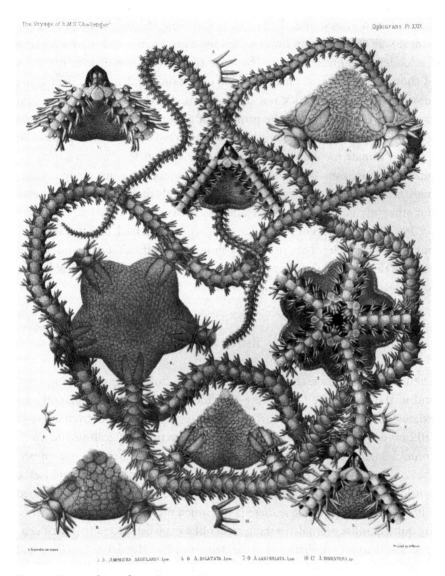

Figure 13 *Amphiura* from Station 50
Source: T. Lyman (1882), Report on Ophiuroidea. Zoology, Part 14, Plate 29. Bound in Reports, vol. 5.

Deserters

Challenger docked the following morning in Halifax, which was cold and snowy even in May. There was the usual run on shore, and indeed some of the sailors ran permanently, never to return. Two seamen and three boys deserted at Halifax; two other boys tried to but were caught and returned to the ship, no doubt to rue the day. This pattern was to be repeated at many of the other ports that the ship visited later: by the end of the voyage, 60 of the original complement of 243 had deserted. A desertion rate of one-quarter of the crew is quite staggering, and shows how hard life before the mast was, even on a cruise where the strict naval discipline of a man-of-war was somewhat relaxed. The eagerness with which the cod were welcomed suggests that the monotony of shipboard meals may have been one reason. Nothing expressed more clearly the difference between officers and men; the men fed on hardtack, pease pudding, and salt horse, while the officers dined almost as well as in port. For

> Officers' wives have puddings and pies,
> But sailors' wives have skilly.

Much later in the voyage, at Amboina, a hunting party of officers recruited some of the local men as beaters. When they paused for lunch, Lord George Campbell explains, "We had brought a large bagful of ship's biscuits for the beaters, on which they feasted, we on German sausage, game pate, bread, claret and whiskey. We offered them some sausage, or rather we offered them the excellent lard in which, inside the tin, it had reposed." On board ship, with officers and men in continual close contact, this unembarrassed assumption of privilege was certain to lead to resentment. A little earlier, indeed, after leaving Tenerife, the officer's meat safe had been broken into and cleared out, despite a marine sentry posted at the door day and night to guard it. It seems unlikely that anyone would take the serious step of deserting ship merely for the occasional rasher of bacon, but the catering arrangements certainly didn't help.

There was little chance of recovering a man once he had run. A reward of £3 was usual, but seldom if ever claimed. There were no passports, identification papers, social insurance cards, health plans, vehicle licenses, cell phones, credit cards, or any of the rest of the paraphernalia that now we carry to ensure our comfort and security, and which preclude any possibility of escaping the attention of the authorities wherever we might be. A seaman in the 1870s could melt away into the crowd to lead any life he chose. The hard monotonous dangerous life on board ship, with no women and no privacy, meant submitting to a discipline that we can scarcely imagine; life on shore, while perhaps equally poor and uncomfortable, brought a freedom that has now slipped away from us.

Challenger left Halifax on 19 May to return to Bermuda before her second, west to east, transect of the Atlantic. Her departure was farcical. She was cheered off by the officers and men of the *Royal Alfred*, whose band struck up "Auld Lang Syne" as *Challenger* steamed away from the hard and prepared to let fall sails. It was at this point that Willemoes-Suhm remembered that he had just sent his servant on shore to fetch a pair of boots, and the man had not yet returned. One can imagine the frown, the pursed lips, the general what-can-you-expect of the Captain and officers. The ship was secured to a buoy and the master-at-arms sent off to shore in a boat to recover the servant. It began to rain. At length, boat, master-at-arms, servant, and boots arrived and the ship could unmoor and make all plain sail south. Its route took it across the western edge of a large flat area of sea floor, about the size of Newfoundland, which is now known as the Suhm Plain. Perhaps some oceanographer was having a little joke.

Station 50: 21 May 1873

42°08' N, 63°39' W; western Suhm Plain, 1,250 fathoms

Sea legs

The dredge was hauled the next day on the banks, with a gravel bottom, and the following day in deeper water on the blue mud of the continental slope. Both hauls gave plenty of work for the naturalists, and both included a very odd-looking kind of animal that they had not seen before on the cruise, a sea spider, or pycnogonid, called *Colossendeis*. This is another group of animals that is obscure without being rare. They look rather like spiders, with eight long legs, and may be distantly and doubtfully related to them. The first question you might ask, though, on seeing one for the first time, is, where is the body? In many species the legs are very long indeed, and the body a mere short stub where they come together: "three-inch length of legs and body comparatively nowhere," as Campbell put it. Their long legs are used to clamber very slowly through the forest of seaweed and animal colonies that encrusts rocks and pilings, pausing to tear off a polyp from a hydroid or a bryozoan and suck its juices. Deep-sea species often have extraordinarily long legs, perhaps to enable them to skim over soft sediment, rather

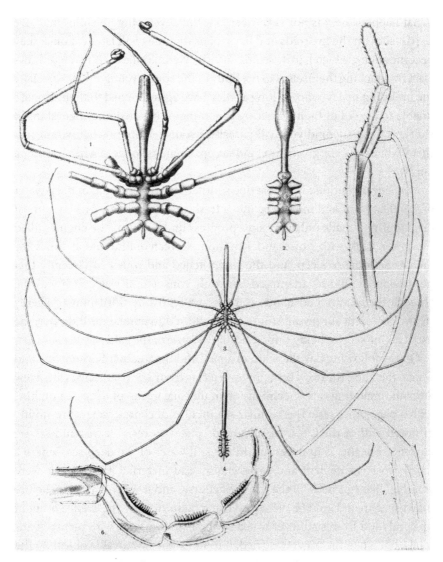

Figure 14 *Colossendeis* from Station 50

Source: Pycnogonid Colossendeis., p. P. C. Hoek (1881), Report on Pycnogonida. Zoology, Part 10, Plate 8. Bound in Reports, vol. 3. Also in Narrative, vol. 1, Part 2 Figure 323.

as a water strider skims over the surface of a pond. They are also used in a much less conventional way: the female stores her eggs in her dilated femurs, or thighs—all eight of them. The eggs are released when she is clasped by a male, who fertilizes them and then glues them together, winds them like a bracelet around his own legs, and broods them until they hatch as tiny larvae.

What happens next is not very clear, but the developing juveniles may live as parasites on the hydroids and bryozoans that they will later eat once they become adults, which hardly seems fair. The victims take their revenge, however, by attaching themselves to the body of the adult pycnogonids, especially the brooding males, who, so long as they have eggs attached to their legs, are unable to get rid of them either by grooming or by shedding their skin. In short, as a pycnogonid you will parasitize your neighbors when young, eat them when you are grown up, and accept them as lodgers when raising a family.

Challenger continued, almost due south, back to Bermuda, in deep water of 2,500 fathoms and more. She met a fresh gale on Sunday, 24 May, almost dead foul, and made only very slow progress under steam. The engines filled the lower deck with noise and vibration, scarcely endurable at times for men attempting to sleep, and the ship pitched and rolled so violently that no dredging could be attempted. The only consolation was that it was the Queen's Birthday, so all hands were issued with one-third pint of sherry (from the "extra surveying stores") and allowed to smoke until dinner, the usual divine service being impracticable because of the gale.

On the following day the wind dropped and the weather became fine and warm, then hot; the trawl brought up red clay from 2,650 fathoms, with a few animals, including rare specimens of an unusual group of deep-sea urchins with a paper-thin test. The familiar sea urchins of coastal waters are stoutly armored with a thick test of calcareous plates to resist waves and strong currents, but this is unnecessary in the still water of the deep sea, where a fragile coat is sufficient and saves energy and material. The animals were smashed like eggshells by the trawl, of course, and it was many years before intact specimens could be recovered using modern gear. The dredge brought up coral mud from shallower water a couple of days later, with plenty of animals, but when the trawl was tried, it found some immovable object on the seabed and was lost, together with 1,700 fathoms of line.

A worm in the wrong place

On 31 May the ship took on a pilot and returned to her former mooring at the jetty in the Dockyard basin. She stayed here for the next 12 days while the naturalists explored the shores and shallow bays, finding lots of animals, but mostly of familiar forms. The only notable discovery was the remarkable worm found by Willemoes-Suhm when he was turning over logs at the edge of the mangrove swamps lining Hungry Bay, where he had gone to watch the land crabs. He found a lot of earthworms, not surprisingly, but also found "long, slimy animals" that he could not immediately identify. When he examined them back on board he discovered that they were not annelids (such as earthworms) at all, but nemerteans. These are unusual animals in which the front part of the body is mostly occupied by a spacious chamber containing a spear attached to a long filament. This can be forcefully ejected to harpoon prey and kill them by injecting a toxin, or more prosaically as a sort of grapnel to haul the animal along in a series of leaps and bounds. Nemerteans are not uncommon and can be found in the intertidal, where one species has a claim to be the longest animal in the world, albeit a very skinny one—up to 55 meters in length, although only a few millimeters wide. A great deal of patience is required to unwind this fragile sticky animal from the rocks and weeds it is wound around. The animals found by Willemoes-Suhm were much smaller, no more than 50 millimeters long, and not particularly unusual, but they were in the *wrong place*.

I know how he felt. One summer afternoon I was helping my grandson turn over logs in Green Park, in the middle of London, when he grabbed a small animal that had scuttled away and almost escaped. The way it moved told me what it was, and a hand lens confirmed the diagnosis: an amphipod, the kind of flattened shrimp that moves on its side. I am familiar with amphipods, but I was astonished nonetheless, because it too was in the wrong place. Amphipods are aquatic animals that live in rivers and lakes and the sea; they do not live in soil in London parks. It was like finding a starfish in a pond. (I found out later that they had been accidentally introduced from New Zealand, where they have such things.) A terrestrial nemertean was equally disconcerting, because nemerteans are exclusively marine: the *Challenger* found them at a score of stations, at all depths in the sea. Only one had ever been found on land before, in the Philippines, although this species and some close relatives have been found subsequently in Australia and New Guinea, and turn out to be widespread in Bermuda. They are a bit of a puzzle: if one, why not many others? There is certainly no impassable bar that prevents nemerteans from living in soil, so why are they not as common and diverse as earthworms or nematodes? Come to that, why don't starfish live in ponds?

The ship took on coal and very welcome fresh biscuit, to replace the old moldy stores that had become infested with maggots and weevils. The naturalists packed the collections made so far for carriage back to London. The gravestone for William Stokes and Adam Ebbels, purchased in Halifax, was erected in the cemetery. Two seamen deserted, despite being on an island from which it was difficult to escape, and a price of £3 was put on their heads (they were never found, and perhaps took a berth on some outgoing ship); at the other end of the social scale the scientists and officers were entertained to a ball thrown by the Governor.

8

Third Leg

The Second North Atlantic Transect, June–July 1873

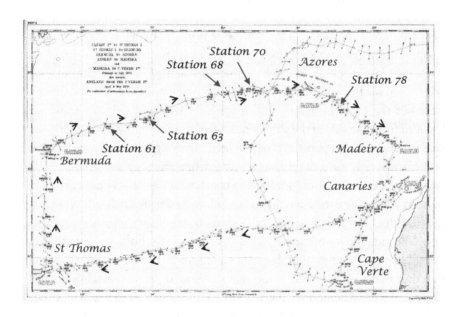

Challenger left Bermuda on 12 June 1873, edging past the "treacherous purple shadows," as Wyville Thomson put it, but lingering in the vicinity of the reefs for a couple of days before Tizard (I think) decided to make a second transit of the Atlantic, and the ship set off north of east for the Azores. Most of the officers and crew went down with dysentery, perhaps picked up from fruit in Bermuda, and severe cases were put on the sick list and treated with a diet of beef tea and arrowroot. All recovered in a few days, and for the next two weeks the ship was pushed along gently by the trade winds, following a course parallel to her first transect but about 15 degrees to the north. She was sailing about 2,500 fathoms above the Suhm Plain, a rather uniform field of red clay larger than France. Its contours are masked by half a kilometer of sediment, part of it dumped by the St. Lawrence, which lies to the northwest, although *Challenger* was passing just to the south of the long chain of

Full Fathom 5000. Graham Bell, Oxford University Press. © Graham Bell 2022.
DOI: 10.1093/oso/9780197541579.003.0009

New England Seamounts, peeping through the clay like pimples, and would shortly sail over a similar cluster further east, the Corner Rise Seamounts, without noticing either.

Station 61: 17 June 1873

34°54′ N, 56°38′ W; Suhm Abyssal Plain, 2,850 fathoms

Few fish had been caught so far, and none from very deep water, which was disappointing. Jellyfish and "insects" were all very well, but no one except the scientifics would set out to catch them deliberately. Fish were different: anyone could see the value of catching new kinds of fish. Fish in the net, though, were not the same as money in the bank. When the dredge or trawl was lowered from the side of the ship, it necessarily sank rapidly through the surface waters, mouth open, and might catch anything it met as it descended. A few small fish were caught on 16 June when the ship sounded in 2,575 fathoms, but there was no mud in the trawl, which probably never reached bottom. A few more were caught on the following day in 2,580 fathoms, when the trawl had certainly reached the bottom, because they were accompanied by ophiuroids and holothurians, which live only on the bottom. Nevertheless, Wyville Thomson knew that they resembled midwater forms and suspected that they had been captured as the gear descended; he was certainly correct. The fish were bristlemouths, named for the thin sharp teeth that fringe their jaws. They rise into surface waters at dusk to feed under cover of darkness on small planktonic crustaceans, and retreat several hundred meters during the day, using a swim bladder to control their buoyancy. The fish caught by the trawl belonged to a genus now called *Cyclothone*, which lives in vast shoals through the oceans of the world, and may well be the most abundant of all vertebrates—by some estimates, there are thousands of bristlemouths for every living person.

Scatter

These rather unimpressive fish—they are dark in color and only about the size of a minnow—have an unexpected connection with antisubmarine warfare. At the time of the *Challenger* voyage there were no ways of detecting or

Figure 15 *Cyclothone* from Station 61
Source: A. Gunther (1887), Report on Deep-Sea Fishes. Zoology, Part 52, Plate 45. Bound in
Reports, vol. 27.

destroying submarines, because there were no submarines, or at least none
capable of traveling submerged for more than an hour or two. The 1870s
were a time of peace on the high seas, when *Challenger* could sail quite safely
around the world after having removed almost all her guns. Forty years
later, in the middle of the First World War, her iron-hulled successors, once
thought invincible, were faced with the novel menace of submarines armed
with torpedoes. Detecting these invisible boats became an urgent military
problem, and toward the end of the war this had been solved by devices that
sent bursts of sound through the water and then detected the echoes reflected by objects in their path. After the war, a similar apparatus was used to
map the sea floor, pointing the sound pulses straight down and timing their
return. This was much easier than casting the heavy deep-sea sounding
lead used by the *Challenger* and gave a much more detailed picture of the
seabed. This picture had one very surprising feature, however: the presence
of thousands of seamounts only a few hundred meters below the surface,
often in areas where such shallow water had never before been suspected.
This was very odd, but each feature was duly recorded on the official charts,
following, I suppose, a precautionary principle that mariners should be
made aware of any possible danger. It became even more odd when many
of these features could not be confirmed by subsequent surveys, and down-
right outrageous when it was found that some of them, at least, rose and
fell several hundred meters on a daily rhythm. What the sonar scans were
showing, in fact, was not the seabed at all, but the shoals of midwater fish
like the bristlemouths, rising and falling to feed and hide, with the sound
waves bouncing off their swim bladders. The biological nature of the "Deep
Scattering Layer" was firmly established by the mid-1940s. It took just a few

years for fishermen to realize that this discovery could be put to good commercial use: the first fish-finder was built in Japan in 1948.

Blink

Bristlemouths live in the twilight zone, between the bright surface waters and the darkness of the deep sea. They are by no means alone, and the most abundant fish that rise and fall in a daily cycle are the myctophids, the lantern fish, whose name gives away their most spectacular feature. Below the twilight zone the sea receives no light from the sun, and yet it is not completely dark. The creatures that live there make their own light, like fireflies in a summer garden. This is not as unlikely as it may seem—the metabolic pathways common to all organisms have a variety of oxidation reactions that can be modified to yield a photon rather than an electron, and the ability to generate a brief flash of light has evolved independently in dozens of quite unrelated lineages of fishes, shrimps, and many other kinds of animal. In imagination, the deep sea is something like the nave of an enormous cathedral on a dark winter's night, with no illumination except that from time to time a candle is lit and then almost immediately snuffed out. These pinpricks of light are continually appearing and disappearing, and something is producing them, but its size and distance are difficult to estimate. An animal in the deep sea floats in a great dark void in which brief lights blink here and there, on and off.

The lights are used to communicate. On the land, on the shore, and in the surface waters of the ocean, animals can use the bright flat light to communicate by color and behavior. In the depths of the sea they must make their own light to communicate with one another. But what are they saying? So far as I know, they are not social animals and are not passing on the location of a particularly succulent patch of plankton or a predator to be avoided. The only information that is likely to be interesting enough for a signaling system to evolve is to declare your species and sex, and wait for offers. The blinks and flashes are sexual signals. Perhaps the cathedral analogy is not quite right, and the deep sea is more like a vast but very sparsely attended nightclub in which the gloom is punctuated by flashes of light as you wander around trying to find a mating partner. You will recognize one by a particular pattern of light and frequency of flashes; this will take some time, because you are unable to see more than 10 meters or so—recall that water absorbs light much more strongly than air—but it is better than simply hoping to bump into someone, and it explains why many deep-sea animals, far from being blind, have large and sensitive eyes.

Signaling by flashlight is not only chancy; it is also very risky, because it provides an opportunity for other fish with big eyes to home in on your signal, not to mate but to eat. They may even have a light signal themselves to lure their prey closer. Some, like the angler fishes, hang out bait, bulbs stuffed with luminous bacteria, to attract shrimps and small fish. Prospective prey can in turn evolve luminous countermeasures, squirting out a cloud of luminous fluid or dropping luminous bombs to deceive predators into attacking a phantom. The twinkling lights of the deep sea, which at first glance seem as innocent as Christmas decorations, are in reality a complex web of deceit, in which the original honest signal of species and sex has become exploited by predators who in turn are misled by their intended prey. Perhaps the nightclub analogy is not so far from the truth.

Station 63: 19 June 1873

15°29′ N, 52°32′ W; Suhm Abyssal Plain, near Corner Rise Seamounts, 2,750 fathoms

Two days later and a little further to the east, just short of the Corner Rise Seamounts, the trawl was put over again and this time captured two kinds of fish. One was undoubtedly from the bottom: a halosaur, a very elongate fish with a long, tapering tail. Halosaurs swim slowly over the seabed, searching the sediment for worms and shrimps. A similar fish had been caught earlier in the cruise, but this was the first indubitable deep-sea fish trawled since beginning the first Atlantic transect, and the scientifics were very pleased—so pleased, according to Matkin, that there was an issue of one-third of a pint of Madeira to all hands. It could hardly have been because of the "large piece of Amber" that Matkin says also came up in the trawl, because no one else mentions it, even the careful Murray. Amber would in any case be an unlikely find in the middle of the Atlantic. Could it instead have been ambergris, the fabulously valuable wax, used as a base for perfumes, formed (of all places) in the bowels of sperm whales? Lumps of ambergris are occasionally trawled from the sea floor or cast up on the shore and found by lucky beachcombers; but no such find is mentioned in the records of the voyage. It would explain the Madeira, though.

Figure 16 *Idiacanthus* from Station 63
Source: A. Gunther (1887), Report on Deep-Sea Fishes. Zoology, Part 52, Plate 52. Bound in
Reports, vol. 27.

Dwarf males

The other fish caught in this trawl was succinctly described by Campbell as
"a long, lean, black ugly fish with a beard." This was a dragonfish, *Idiacanthus*.
It is one of the deceitful luminous fish that prey on the swarms of midwater
plankton-eaters like *Cyclothone*. It is indeed elongate, and this is emphasized
by its enormous head and fearsome fangs that can catch, subdue, and swallow
anything that it encounters. It is a living trap, with an outsize mouth and stomach because meals are few and far between, and it must make the most of
any prey it comes across, large or small. The "beard" is a luminous lure to
attract the prey. The animal that was intercepted by the trawl—in midwater,
certainly—was a female, a foot or so in length; the males are much smaller, a
couple of inches or so, have no fangs and, oddly enough, lack a beard. Their
mission is not to feed, but to find a female. This is quite a common pattern in
animals, and reflects the fundamental inequality between a large nutritious
egg, fueled to feed the early development of the embryo, and a minute sperm.
Females become specialized for provisioning the embryo, while males dash
around to fertilize as many females as possible.

Another animal in the trawl illustrated this principle in an even more dramatic form. *Scalpellum* is a stalked barnacle that attaches itself by a stout pe-
duncle to a pebble or lump of clay on the seabed, so there is no doubt that
the specimens had been brought up from the bottom. It is a stout, more
or less bean-shaped animal a couple of inches long; it looks a little like a
clam, enclosed in two shells, but the animal inside is more closely related to
shrimps, protected by its shell while waving its legs in the water to strain out
algae. At least, that is what the females do. The males are minute, scarcely
visible, and attach themselves to a fold in the body wall of the female. They
are incapable of feeding—they have neither mouth nor gut—and rely on the
female to furnish their nutrition. Instead, their bodies are crammed with
sperm. They have become the ultimate in sexual inequality, a swimming

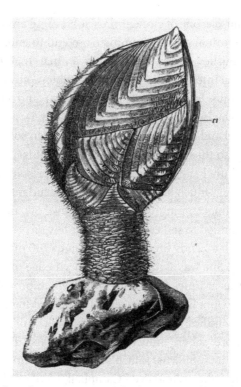

Figure 17 *Scalpellum* from Station 63. The large animal attached to a nodule is female: the dwarf male clinging to her is indicated by *a*.
Source: Narrative, vol. 1, Part 2 Figure 321.

testis that settles on the female as a sexual parasite and eventually becomes little more than an accessory organ.

The ship was now moving away from the abyssal plain over a more complex region of ridges and canyons formed by faulting at right angles to the Mid-Atlantic Ridge, although the depth recorded by the sounding lead remained steady at around 2,700 fathoms. On the following day the dredge brought up a few animals, including the only priapulid recorded during the voyage. Their common name is said to be "penis worms" from their resemblance to a large flaccid penis, although I strongly suspect that they do not really have a common name because they are discussed solely by zoologists, who always refer to them as "priapulids." In any case, the specimen in the dredge was only 16 millimeters long by three millimeters wide and "very much spoiled," according to Willemoes-Suhm. Priapulids have a particularly simple construction, consisting basically of a cylinder of seawater enclosed by two layers of muscle, one of circular muscle running around the body and the other of longitudinal muscle running from end to end. Contracting the circular muscles pushes out the front of the animal, making it longer and thinner, because the internal fluid is incompressible; conversely, contracting the longitudinal muscles makes the body shorter and fatter. Alternately contracting and relaxing the two sets of muscles rhythmically changes the shape of the animal and enables it to make a burrow in soft sediment, where it lies in wait to grab the occasional passing worm. It is a simple way of life that has remained unchanged over the ages: fossil priapulids, in their fossil burrows, looking very much like their modern descendants, are known from the very earliest animal communities, that lived over 500 million years ago.

Challenger continued to sail east for the next few days toward the Azores, in fine weather with a handy breeze, making up to 200 miles per day, over a bottom of red clay. She was skirting the northern edge of the Sargasso Sea, and the surface was at times thick with weed, which provided Murray and Moseley with diversion. Everything seemed to be shaking down nicely. One can picture the scientifics: the jovial, very slightly comic, figure of Wyville Thomson, the more sober, industrious Murray and Buchanan, earnest Willemoes-Suhm, the somewhat raffish Moseley. All seemed to be getting on splendidly with the officers, and even Campbell, who is sometimes inclined to be a little cynical about the value of natural history, now professed some interest in the contents of the trawl. The crew were not much mentioned, but can be imagined to be working in good heart with the dredge and sounding gear and larking in the sunshine. Perhaps it was so; but Swire gives a different picture. At about this time, he records in his journal:

> Sailors are not jolly fellows by any means, as a body, neither officers nor men; on the contrary, they are very disagreeable, and only show jollity in the occasional perpetration of objectionable practical jokes, and as for heartiness, if an unfortunate disability to say necessarily disagreeable things in a polite way be a merit, sailors may be considered meritorious; but rudeness is not heartiness, and most people who come into contact with sailors put them down as a rude and boorish class. . . . Can the pleasure of visiting foreign countries and of breathing fresh air (and that only as long as you stop on the upper deck) atone for the loss of education entailed by entering the Navy, the monotony of life at sea, the petty tyranny to which the sailor is subjected, the total estrangement from all home ties, and the fact of being shipmated with a set of people for whom one does not care a brass farthing,

and who do not care a brass farthing for you, and of one and all of whom you are heartily sick before the first year of a commission has elapsed?

Bear in mind that this is not an opinion piece in the *Guardian*, but an extract from the private journal of a serving officer in the Royal Navy. It certainly gives a picture of life on board ship quite different from what you might imagine from reading Wyville Thomson or Spry or Moseley. Swire's impressions of the individual scientifics and officers would be very interesting to read, but one leafs through his journal in vain: before publication (according to the foreword), "He very carefully (but very completely) obliterated all passages which he thought could in any way give offence to any living person or be interpreted as casting the slightest slur on the memory of anyone who had died." Which implies that there were lots of them.

Station 68: 24 June 1873

38°03' N, 39°19' W; southern end of Newfoundland Basin, 2,175 fathoms

There is an unexpected link between medical technology and the ecology of the deep sea. Machines that count particles such as red blood cells automatically were invented in the 1960s and were soon used in a range of medical and industrial applications. It was not long before R. W. Sheldon and his colleagues at the Bedford Institute in Nova Scotia took an automated particle counter on board ship and ran seawater through it to measure the range of size of particles in the ocean. More precisely, they wanted to know what fraction of the total mass of suspended solids is made up of particles within a given size range. I don't know why they did this or what they expected to find, but they arrived at a very simple conclusion, that the frequency of particles is inversely proportional to their mass. This rule holds whether or not the particles are living organisms. Hence, it implies that in any sample of seawater the combined mass of organisms within some given range of size is a constant. The smallest organisms are bacteria, which are very small and very numerous; almost every organism between 10^{-12}g (a million-millionth of a gram) and 10^{-10}g (100 times bigger) is a bacterium, and there are lots of them. At the other extreme, large sea creatures such as fish, jellyfish, starfish, and squid mostly weigh between 10 grams and 1 kilograms, and there are far fewer of them. Sheldon's rule tells us that larger size is balanced by lower abundance, so the combined mass of the bacteria is equal to the combined mass of the megafauna. The largest marine animals—mostly whales—weigh between 1,000 kilograms (a minke whale, say) and 100,000 kilograms (a blue whale) and will have the same combined mass in a given volume of ocean, or at least would have before the first long-distance whaling boats set out from Nantucket. The mass equivalence of bacteria, starfish, and whales is not always exact by any means, but shows how an astonishingly simple generalization can emerge from new technology that was developed for some quite unrelated purpose.

Weedfall

What does this have to do with the deep sea floor? Well, all these organisms, large and small, will sooner or later die and sink, and as they sink they will decompose. When they begin their descent from the surface, Sheldon's rule tells us that large and small will be equal in combined mass. Smaller objects sink more slowly, however, so smaller bodies will be more thoroughly decomposed at any given depth. A single algal cell will sink at less than one meter per day, and would take five years to sink 1,000 fathoms, except that it would never get that deep because by this time it would have completely disintegrated. Larger objects sink faster, and much larger objects sink much faster: a hand-sized clump of seaweed might sink at 50–200 meters per day, a large jellyfish at 1,500 meters per day, and a dead whale, once the sharks had finished with it, would reach the sea floor, no matter how deep, on the day it died. Because they sink faster, larger bodies will be less completely consumed, and will make up a larger fraction of the edible tissue reaching the bottom. In deep water, the harvest that the sea floor receives from the surface will be dominated by these rare windfalls—or more precisely weedfalls, jellyfalls, and whalefalls. This is why a distinct community of deep-sea scavengers has evolved to take advantage of these rare but extremely profitable opportunities. They are not sedentary creatures like crinoids, patiently sieving the thin gruel of seawater at depth, but are instead actively moving animals that depend on being able to detect a fall from far away, get to it quickly, eat it fast, and consume it as completely as possible. This characteristic community of spot-and-dash scavengers includes—generally in order of appearance— hagfish, bony fish such as spiny eels and rat-tails, decapod shrimps, giant amphipods, isopods, and, lumbering along in the rear, ophiuroids.

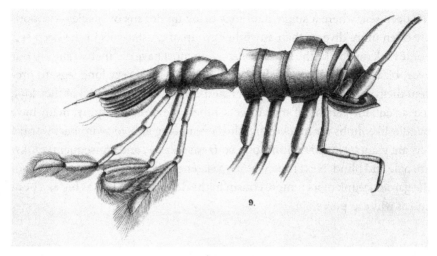

Figure 18 *Eurycope* from Station 68
Source: F. E. Beddard (1886), Report on Isopoda. Zoology, Part 48, Plate 14. Bound in Reports, vol. 17.

Slaters

These scavengers can be studied by mounting a camera on a parcel of bait placed on the seabed. Tuna or mackerel baits quickly attract fish and amphipods; seaweed and spinach are less attractive (parents will have noticed this pattern too), but seem to be relished by isopods, which have been seen to tear off pieces of *Sargassum* and retreat with them into the gloom. Isopods are the only group of crustaceans that has successfully adapted to life on land; they are the familiar wood lice, pillbugs, and slaters. Aquatic crustaceans, the great majority, breathe through gills borne on their legs. These are useless in air because surface tension makes them collapse, but in slaters they have become covered with a thin outer wall and converted into a kind of lung. This works well enough in damp places, where slaters lead blameless if unexciting lives eating leaf litter. Aquatic isopods also eat decaying vegetable matter, trundling slowly across the mud at the bottom of ponds or the sea. The bed of the Sargasso Sea is littered with sunken fragments of gulf-weed whose air bladders have been punctured by storms or grazing animals, and these are cleaned up by isopods. Between treats like this they probably search through the clay for edible fragments, but not much is known about their feeding habits, or indeed about any of their other habits. They are not a charismatic group. They are, however, a remarkably successful group in the deep sea, where a single haul may bring up dozens of species—isopods are often more diverse than any other group of crustaceans in the deep sea. Some look much like the familiar slater, but most have distinctive and (to our eyes) bizarre adaptations to deep-sea life: some have very long legs, to prevent them sinking into the sediment, and elongated bodies, so that they look like an aquatic version of stick insects; others are stout and spiny; many have paddle-like limbs for maneuverability, even being able to swim backward; a few are giants. None are luminous, so far as I know, and consequently most are pale and blind. Next time you see a slater creep away in a damp corner of the garden, think of its remote cousin in the depths of the sea, as big as a boot and as white as porcelain.

A frangible bag

The following day, sounding in 2,200 fathoms, the trawl brought up an extraordinary object. Even Campbell was impressed. It was a great bag-like animal consisting of tens of thousands of small pink lumps; it was alive, and when placed in a tub on deck it shone brightly all night. The officers traced their names on its surface, and saw them appear in glowing lines. This was *Pyrosoma*, a colonial tunicate. Tunicates are rather distant relatives of vertebrates like ourselves that have retained the ancestral way of life. Most of them are rather stodgy creatures about the size of a fist that can be seen attached to harbor walls or dock pilings. They make a living by filtering seawater: water is drawn in through a siphon, forced through slits in the wall of their throat to filter out the plankton, and then expelled through a second siphon. In fish these slits have become gill slits, and the supporting structures of the most anterior slits, near the mouth, have evolved into jaws. They still appear transiently in human development, a reminder that evolution seldom innovates but rather modifies preexisting structures. Some tunicates have left the harbor wall and use the water jet from the exhalent siphon to propel themselves through the water, feeding on the move. *Pyrosoma* is a colonial swimming tunicate, in which the combined water jets of thousands of miniature animals, all budded from a single progenitor and bound together in a gelatinous envelope, push their family group through the water. Murray thought that the trawl had never reached bottom, and he was undoubtedly right; *Pyrosoma* is a midwater animal that was probably captured quite near to the surface. The following morning the tub seemed at first sight to be empty with the animal gone, but in fact it had simply shattered into thousands of separate little pink lumps, the individual members of the colony. The previous evening, someone had the idea of trying an electric shock, which had no immediate effect but might have distressed the animal; as we now know, severe stimulation often causes the colony to fall apart.

Station 70: 26 June 1873

38°25′ N, 35°50′ W; Pico Fracture Zone, 1,675 fathoms

Challenger was now sailing in fine weather over a rough topography of ridges and valleys as the ocean shallowed on the approach to the Mid-Atlantic Ridge and the red clay gave way to ooze. The trawl brought up some of the ooze and some slimy brown lumps that afterward turned out to be much more interesting than they seemed to be at the time.

Haeckel's mistake

Ernst Haeckel was a very well-known German zoologist of the late nineteenth century, one of the earliest and most influential supporters of Darwin. He wrote several of the Reports that later described the animals found by the *Challenger*, including his account of the radiolarians, which has become famous, even outside radiolarian circles, for its stunningly beautiful

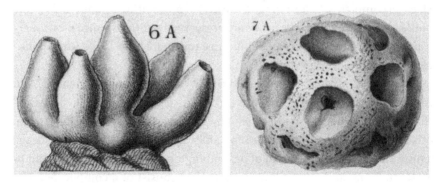

Figure 19 Xenophyophores from Station 70
Source: E. Haeckel (1889), Report on Deep-Sea Keratosa. Zoology, Part 82, Plate 7. Bound in Reports, vol. 32.

illustrations. A much less aesthetic task was presented by a few dozen bottles containing small brown lumps floating in cloudy alcohol. He identified them as sponges, but from the outset his report has a decidedly defensive tone. This is odd. Sponges constitute a phylum of animals. A phylum is a traditional group whose members share a common body plan, usually with some unique feature that serves to identify it, such as mollusks (distinguished by the radula, a scraping structure used in feeding), arthropods (jointed limbs), cnidarians (stinging cells), and ctenophores (comb rows, used in locomotion). Any competent zoologist presented with an animal collected at sea, in a lake, or on land can unhesitatingly assign it to a phylum—well, there are a few exceptions, but membership of a phylum is seldom controversial. Sponges are filter feeders, and their unique distinguishing feature is the flagellated chambers that drive a water current through their body. Haeckel could not find any flagellated chambers, and blamed the poor preservation of the specimens. He nevertheless identified them as sponges—what else could they be?—but he faced opposition from other *Challenger* authors. The report on the Keratosa (the group of sponges to which Haeckel assigned his specimens—nowadays we would call them Demospongia) was written by N. Poléjaeff of Odessa, who begins his account with the categorical statement, "The Keratose Sponges do not belong to the deep-sea fauna," a conclusion that Haeckel strongly disagreed with, since some of his material had been collected below 2,500 fathoms. More troubling, Henry Brady described very similar specimens (as Haeckel acknowledged) in his report on the foraminiferans, and interpreted them as gigantic bottom-dwelling relatives of the minute surface-dwelling rhizopods whose sunken remains form the globigerina ooze. This must have been an old difference of opinion, because Haeckel complains, "I find in his [Murray's] handwriting on the labels of the bottles in which all the large forms are preserved the title "Sponges," but afterward another naturalist crossed this name out and wrote "Large Rhizopods." How dare they!

Haeckel was wrong, however, and Brady (and whatever unfriendly hand that crossed out Murray's label) was right. The specimens are indeed gigantic rhizopods, not sponges at all, which now go by the almost unspellable name of xenophyophorans. Most are rounded structures of deeply folded sheets, a little like a crumpled-up ball of paper, ranging in size from a small coin to a saucer. Some look rather like a brain. They consist of a complicated network of branching tubes in which the nuclei are scattered without distinct cell boundaries, a sort of boulder of cytoplasm. The "xeno" part of their name reflects their habit of collecting all sorts of foreign objects—sand grains, sponge spicules, their own feces—and gluing them together to make a loose covering, or investment, that usually falls apart when they are roughly handled by being plucked from the seabed by a dredge. They live in all oceans, generally in the abyssal plains, and are often abundant, especially in canyons and on the sides of ridges. They may very well be ecologically important in deep-sea communities, but they are seldom studied. Haeckel was wrong, but, over a century later, we know little more about these curious organisms.

Challenger was now moving upslope toward the crest of the Mid-Atlantic Ridge, which she reached on 30 June 1874, just missing one of the most remarkable features of the ocean floor. The Ridge is punctuated by cracks in the sea floor from which plumes of seawater, heated by contact with the hot magma deep below, rise to the surface of the sediment. This hydrothermal fluid is hot, acid, and anoxic, and contains a variety of chemicals such as sulfides, methane, hydrogen, and heavy metals leached from the hot rock. It is an extremely hostile environment, but offers some unique opportunities because substances such as methane, hydrogen, and hydrogen sulfide can be oxidized to yield energy that can be used as the basis for metabolism. Many bacteria can perform these chemical reactions, so that hydrothermal vents support a rich microbial community that uses dissolved chemicals as an energy source, much as the phytoplankton at the surface of the sea uses sunlight to drive photosynthesis. Animals cannot use the hydrothermal fluid directly—animals have much simpler and less imaginative metabolic systems than bacteria—but they still manage to live near the vents by farming the bacteria. Deep-sea animals such as clams, scale worms, and barnacles have evolved the ability to culture bacteria on their surface or even inside their bodies, and make a living by cropping them or using the organic substances they secrete. In many of these animals their normal feeding structures have atrophied to the point where they are completely dependent on their bacterial farms. The result is an extraordinary assemblage of animals able to live at high temperature with very little oxygen, found only on and around hydrothermal vents on mid-ocean ridges throughout the world.

Absence of a gutless worm

The most extraordinary of all are the peculiar tube worms called pogonophorans, which were not discovered until after the *Challenger* voyage.

When they were discovered, by a Dutch expedition dredging off Indonesia in 1899, they puzzled the zoologists because they lacked any trace of a digestive system—no mouth, gut, or anus could be made out. Perhaps the specimens were broken, and the rest of the animal had not yet been found? In time it was realized that they really do lack a digestive system, raising the difficult problem of how they feed. Perhaps the tentacles borne on the head could perform some sort of extracellular digestion? Or perhaps the whole body could absorb sugars and amino acids from the sediment? The correct answer lay in a unique structure in the posterior part of the trunk, formed by the degeneration of the larval gut, which turns out to be packed with bacteria. Pogonophorans are bacteria farmers. Naturally, this is a way of life ideal for living on hydrothermal vents, and the vent tube worms are gigantic animals up to a couple of meters in length, packed with bacteria that oxidize the toxic plume of hydrogen sulfide that rises from the sea floor.

Challenger passed just a few hours' sail northwest of the Menez Gwen and Lucky Strike ventfields, and as far as I can make out never did sample the vent fauna, either here or at later near misses. It would be pleasant to record that the trawl brought up gigantic inexplicable worms as long as a man, but it never did. In fact the naturalists never did report anything that we could now interpret as a pogonophoran, although they may have caught some unawares; the broken tubes of the run-of-the-mill relatives of the vent tube worms look much like frayed threads of line or net, and might have been overlooked.

The ship sailed on through a sea that shone brilliantly at night with luminous salps, siphonophores, and jellyfish, heading for the Azores. These are volcanic islands rising from an underwater plateau in the middle of the ocean at the triple junction of the Eurasian, African, and North American tectonic plates. They used to be shunned by mariners because the prevailing westerlies make them a lee shore, with few good natural harbors. The islands were sighted on the morning of 1 July, and, pushing slowly through patchy mist, *Challenger* anchored just off Horta, the main town of the island of Fayal. A local official rowed out to enquire minutely into the health of the crew and of the ports at which the ship had called recently, quite routine, except that he forgot to mention that there was, in fact, an outbreak of smallpox in the town itself. When they were told of this by the British vice-consul, Nares wisely decided to move on, but rather unwisely went on shore himself with some of the scientifics. Moseley appears simply to have jumped ship: "I slipped on shore in a fruit boat, or I should not have been

allowed to land at all." Fortunately, the smallpox did not take advantage
of this outrageously irresponsible jaunt. The ship left Fayal the following
morning and messed around in the shallow channel between the islands
of Fayal and Pico for a couple of days, dredging up pumice stone and hundreds of animals, before setting off for San Miguel, the largest island of the
group, and anchoring off Ponta Delgarda on 4 July. The men were allowed
on shore, where they got fighting drunk, while the naturalists visited the
hot springs of the interior. All eventually returned, some slightly battered
(one seaman had somehow ripped the seat right off his trousers, and had to
turn continually from side to side to conceal this from the officers), and the
ship left San Miguel on the evening of 9 July, setting a course for Madeira,
which lay some 600 miles to the southeast.

Station 78: 10 July 1873

27°26′ N, 25°13′ W; off San Miguel, Azores, 1,000 fathoms

The seabed off the coast of the Azores is buckled into a jumble of seamounts,
ridges, and pits by the forces that are pushing the continents apart. *Challenger*
set off southward through this underwater landscape, passing over a small
unnamed basin, floored with volcanic detritus, between San Miguel and the
tiny island of Santa Maria. The dredge brought up a lot of mud and a lot of
animals—sponges, corals, echinoderms, and crustaceans of all kinds—including about 30 species that had never before been described. It was still
possible in those days to find 30 unknown animals, belonging to species that
nobody had ever seen before, in a single haul of the dredge.

A cup of coral

The most conspicuous were the corals. These were the genuine article, stony
corals encased in calcium carbonate, rather than the soft alcyonarian corals
such as sea whips and sea fans. They were not reef corals, though. Reef corals
are colonies made up of thousands of minute polyps that depend on the symbiotic algae that they farm in their tissues to supply much of their nutrition.
These corals were much larger, solitary polyps such as *Fungiacyathus*, lacking
symbiotic algae because they lived far beyond the last gleam of sunlight, and

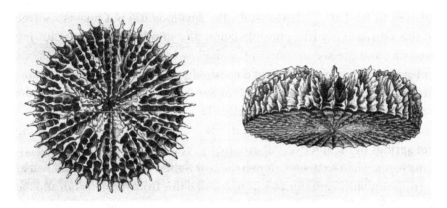

Figure 20 *Fungiacyathus* from Station 78
Source: H. N. Moseley (1881), Report on certain Hydroid, Alcyonarian and Madreporian Corals.
Zoology, Part 7. Bound in Reports, vol. 2 Figure, p. 186. Also in Narrative, vol. 1, Part 2 Figure 287.

therefore entirely dependent on their stinging cells to catch living prey such as small crustaceans. You will not be far off if you think of them as being about the size and shape of a small ice-cream cone. The cone is made of calcium carbonate (although in deepwater species it is often quite delicate and fragile) buttressed with internal ridges. The animal inside is the ice cream, squidged down into the furthermost recesses of the cone, as it should be. The closed end of the cone is anchored in the sediment, while the open end bears tentacles armed with stinging cells that capture any small creature that blunders into them. The dredge caught these too: small elongate isopods, tiny comma-shaped cumaceans, and the even smaller blind tanaids, bearing claws like miniature lobsters. All of these are numerous and diverse in the deep sea, and probably quite important in its economy, as they continuously work and rework the sediment, but even now very little is known about their habits.

Challenger continued southwest, passing between Santa Maria and Formigas Hill, which rises to within 25 fathoms of the surface, and skirting Formigas Hole, which plunges 2,500 fathoms beneath. She was now passing over deep water floored with globigerina ooze, until the sea began to shallow on the Madeira Rise. *Challenger* anchored off Funchal in the evening of 15 July. Everyone was looking forward to Madeira, but they were disappointed, for there was smallpox in the town. The crew were not allowed to go onshore, of course, but Nares invited the cream of Funchal society on board and took them for a trip to the Desert Islands, a few hours' sail to the east, from which they soon returned, the ladies being seasick, and instead fell to dancing on the upper deck. The crew had an issue of wine. The party broke up later in the afternoon, and *Challenger* departed in the evening, heading south for the Cape Verde Islands. No smallpox appeared; perhaps the upper classes really were immune.

9

Fourth Leg

Into the South Atlantic, July–September 1873

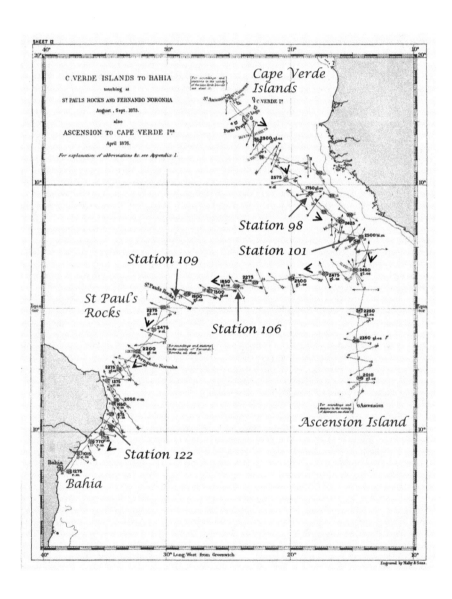

Full Fathom 5000. Graham Bell, Oxford University Press. © Graham Bell 2022.
DOI: 10.1093/oso/9780197541579.003.0010

Challenger found the northeast trade winds and started her run south, skirting the coat of Morocco. She crossed the Canary Basin in a day and dredged in volcanic mud off the island of La Palma on 19 July. Moseley was surprised to find a holly leaf in the mud, the first fragment of land vegetation to appear. The dredge also yielded a good haul of animals, including dozens of mollusks, bivalves, and snails. The most distinctive were the scaphopods, or tusk shells. These are uncommon in shallow water, although you will occasionally find their shells cast up on the shore. They are quite uniform in shape and unmistakable: a slightly curved narrow cone, much like a tusk, open at both ends. They live with their head (at the wider end of the shell) permanently buried in the sediment and the narrow end poking out, for ventilation. The head bears a peculiar arrangement of fine filaments that continually probe through the sediment, searching for diatoms and forams. They are much more common in the deep sea than in shallow water, and turned up in a couple of dozen dredges during the voyage, down to 2,600 fathoms.

One question that had been nagging the scientifics were the unexpected manganese formations, which they first found coating dead coral the previous February. They were now back in the same part of the ocean, and spent a couple of days looking for the same patch of rocky ground, to see whether they could recover any live coral. They found it on 21 July, and dredged up more coral, but it was all dead, and identical to what they had found before. The mystery continued.

Station 89: 23 July 1873

22°18′ N, 22°02′ W; edge of Canary Basin, 2,400 fathoms

The ship continued to sail in fine weather parallel to the coast of Africa, which lay some 300 miles to the east, close enough for a beetle to be blown aboard one day. The seabed on the outer rim of the continental slope was a more or less uniform plain of ooze, some 2,300 fathoms deep.

The sea devil

On 23 July the trawl came up with an extraordinary animal, a small but grotesque fish known (to the small circle of people who chat about this sort of

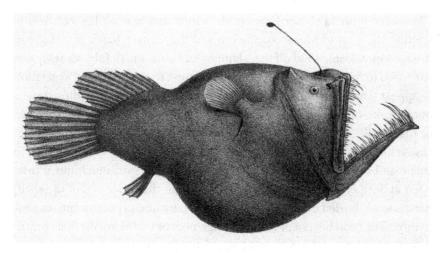

Figure 21 *Ceratias* from Station 89

Source: A. Gunther (1887), Report on Deep-Sea Fishes. Zoology, Part 52, Plate 11. Bound in Reports, vol. 27.

thing) as the stargazing sea devil, *Ceratias*. It had the huge head typical of angler fishes, such as the gulfweed angler of the Sargasso Sea, and a complete set of fishing equipment. The rod and line are formed from the first spine of the dorsal fin, which bears a bulbous luminescent lure at its tip. The lure is quite a complex organ. It consists of a light gland enclosed in a light-proof cup that is lined with a reflecting layer and has a narrow pore that connects it with the surrounding seawater. The gland itself is a nest of chambers containing luminescent bacteria that are presumably recruited from seawater via the pore and afterward nourished in some unknown manner by the fish. The blue-green light emitted by the bacteria is beamed to a transparent window at the tip of the lure through a tubular light guide. Any prey attracted to the lure are captured by suction, which is why angler fishes all have big heads with gaping distensible mouths. Every species of deep-sea angler—and there over 300 of them—has its own distinctive lure that differs from others in details of size, presence of tassels or appendages, structure of light guide, length of rod, and so forth. If there were only one type of lure that all these species displayed, their prey would soon evolve to ignore it, but with so many different kinds of lure dangled before them they cannot ignore them all without losing their own ability to communicate. The diversity of lures is one example of the advantage of being rare, to forestall recognition by potential predators or prey.

Wyville Thomson thought there was no reasonable doubt that the fish lived on the bottom, and imagined it lying buried in the mud with only rod and lure (which he interpreted as sensory structures) exposed, waiting for its prey. In fact, it is a pelagic fish that swims slowly along in deep water like a moving trap, sucking in any shrimp or small fish attracted by its lure. Or, rather, that is what the females do; this is another animal with dwarf males, like the dragonfish and deepwater barnacles. In some species the sexual dimorphism is so pronounced that the female is a million times larger than the male; if our species had the same system, a man would be about the size of a guppy, firmly embedded in the belly of a woman.

Von Willemoes-Suhm also noticed a small transparent worm-shaped animal, barely half an inch long, even more firmly attached to the fish. This was not a male, but an anchor-worm, *Lernaea*, a common parasite of fish that attracts more attention nowadays than it used to because it can damage aquaculture stocks. The puzzle is to work out what kind of animal it is. The

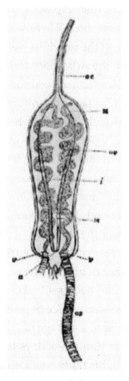

Figure 22 *Lernaea* (parasitic on *Ceratias*) from Station 89
Source: G. S. Brady (1883), Report on Copepoda. Zoology, Part 23, Plate 4. Bound in Reports, vol. 8.

adults consist of little more than the anchor, an X-shaped structure sunk in the flesh of its host, a long tubular body with an intestine, and a pair of egg sacs. It does not much resemble anything familiar, but it is in fact not a worm at all, but a crustacean, closely related to the free-living copepods that swarm at the surface of the sea, that has been so strongly modified by its parasitic way of life that it is no longer recognizable. Its true nature is given away by its larva, which lives in the plankton like other copepod larvae and has therefore remained much the same. The course of development is often more conservative than the final adult form, and sometimes enables us to interpret animals that have been so extensively modified by adaptation to some unusual way of life that they are no longer recognizable as adults.

Over the next few days the sea became shallower as the ship approached the archipelago of Cape Verde. She anchored off Porto Grande, St Vincent (Sao Vicente), on 27 July and stayed there a few days to coal (250 tons costing £5 a ton, nearly 10 times the price at the pithead in England). Deprived of the delights of Madeira, the crew were distinctly grumpy about Cape Verde. "St Vincent is, I think, the most dismal place I know," complained Campbell. There were not even fresh vegetables. There was also no schoolmaster. Mr. Briant had arrived on the island the previous week, having been sent at short notice from Southampton by the Admiralty to intercept the *Challenger* as a replacement for Adam Ebbels, who had died the previous April in Bermuda. The notice had been so short, in fact, that nobody had thought to pay him, and he arrived on the island with just £2 in his pocket, enough to live on for two or three days. Not to worry: he took the natural recourse of a distressed British subject and went for help to the British Consul, a Mr. Graham, who turned him down flat, denying any responsibility and declining to provide any aid. He was now more distressed than ever, but calculated (he kept a journal, which is how we know) that £2 would pay for a week at the hotel he was staying in, leaving only the problem of food. That afternoon he set out for a walk, and never returned. Search parties were sent out from the ship, but failed to find any trace of him in that hot, dry, rocky landscape. The general opinion was that he had been robbed and murdered, especially as a man known to be a murderer—his choice of profession did not seem to bear any severe social stigma—lived in a cottage nearby. He was given up for lost, and *Challenger* sailed without him a few days later, on 5 August. In fact he had not been murdered at all; his body was found a week or so later on the side of a hill a few miles out of town, with his watch and purse intact. He had strayed a little too far, and died, I suppose, of thirst and heat and despair. As it turned out, he had investments in England worth several hundred pounds, and could readily have repaid a small loan.

The unfortunate Mr. Briant arrived on the same boat as a new sublieutenant, H. C. Harston, replacing Sublieutenant Slugett, who had left the ship at Halifax. He was likewise short of money when he arrived, and likewise petitioned Consul Graham, who invited him in and provided room and board until *Challenger* turned up. He was, after all, an officer and a gentleman.

Station 98: 14 August 1873

09°21′ N, 18°28′ W; Kane Passage, 1,750 fathoms

The ship moved from this depressing town to Porto Praya (Praia), on the nearby island of Santiago, where she anchored with some difficulty, because the harbor, although protected from the prevailing westerlies, still caught a long Atlantic swell that rolled the ship around. Once there, however, everyone found this a more attractive town, and there were plenty of fruit and vegetables for sale—bananas, pineapples, mangos, and coconuts. The naturalists dredged for precious red coral from the steam pinnace; the officers (and Moseley, of course) shot quail, and the men went fishing. The boatswain, Richard Cox, conducted the fishing, as usual, aided by 30 or 40 men and boys. They paid out a seine net in a semicircle from the shore, and then hauled it in, standing up to their necks in the sea, until the cod end, filled with fish, could be drawn up onto the beach. This haul was even more exciting than usual, because the catch included a fourteen-foot shark, which Cox prevented from bursting the net by hammering it on the head with a boathook whenever it made an effort to escape the narrowing arc, so that it was eventually pulled up clear of the water, where the sailors dealt with it in the usual way.

Challenger left Santiago on 9 August and set off southward across a broad bowl some 400 miles long, whose rim is floored with volcanic mud, which gives way to ooze and finally to red clay in its deepest part, at below 2,500 fathoms. She had crossed the Tropic of Cancer two weeks ago, to the north of the Cape Verdes, and was now fairly into the tropics. There had even been monkeys and parrots at Santiago; one of the ship's boys had bought a monkey and released it into the rigging, but monkeys are not allowed on board until the ship is homeward bound, so he was made to chase it down and toss it overboard.

The ship was running with the Guinea Current, which branches off the main clockwise circulation of the North Atlantic to run south then east around the bulge of Africa. The sea became brilliantly luminous at night, at first with the constellations of pyrosomes, each as large as a man's hand, and a little further south with the more even, sustained light produced by a single-celled planktonic creature, a dinoflagellate, called *Pyrocystis*. In the perpetual darkness of the deep sea, luminescence evolves as a sexual signal, but this does not seem to be a very likely reason for surface-dwelling organisms to be luminous. My favorite explanation is the "burglar alarm" theory: a dinoflagellate that feels a jolt takes this as a sign that a copepod intending to eat it is nearby, so switches on a light to attract a fish that will eat the copepod. I fear that there is not as yet conclusive proof of this curious and elegant idea.

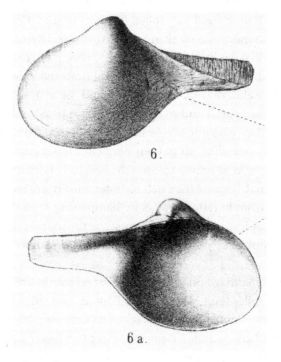

Figure 23 *Rhinoclama* from Station 98

Source: E. A. Smith (1885), Report on Lamellibranchiata. Zoology, Part 35, Plate 10. Bound in Reports, vol. 13.

A fierce clam

On 14 August the ship was passing through a narrow throat between the Guinea Terrace to the east and the extensively faulted zone to the west, at the eastern edge of the McGowan group of seamounts, a swarm of underwater mountains some of which rise to within 100 fathoms or so of the surface. These are incipient volcanoes formed within the African plate, perhaps by the irruption of a deep mantle plume; the *Challenger*, passing about 15 miles north of Jane Seamount, did not detect them. The dredge was lowered here, with disappointing results: Wyville Thomson commented, despondently, "The dredging was not very successful . . . beyond some fragments of a sponge, a broken sea-egg [sea urchin], and one or two bivalve-shells, the dredge contained no examples of the larger animal forms." One of the clams, however, although admittedly not much to look at, a fragile, transparent shell the size of your little fingernail, turned out later to have a remarkable way of life—remarkable for a clam, at least. Most bivalves are filter feeders, pulling in a current of water through their inhalant siphon, straining it through their gills, and eating whatever phytoplankton sticks, then expelling the water through a second, exhalent siphon. The force driving the water current is produced by the beating of thousands of tiny hairlike cilia, which has always seemed to me a rather low-tech solution. Many animals, after all, use muscle power to generate much greater force to drive blood around the body, and it is not obvious why clams should not use a muscular pump to drive water through their gill. It is certainly not impossible, because the animal caught by the dredge (now called *Rhinoclama*) is one of the very few bivalves that can do this: it has a septum, a sheet of muscle something like the diaphragm of a mammal like us, which contracts rhythmically to draw a stream of water into its mantle cavity and over the gill. So far, so good; but now, having evolved a more powerful respiratory system, it has gone further and weaponized it. Rather than graze on phytoplankton, it captures and eats single large prey (well, fairly large—small crustaceans or their larvae, a millimeter or so in size) by using its inhalant siphon like a vacuum cleaner hose. The tip of the siphon bears a set of sensitive tentacles; if these detect movement within a couple of millimeters, the siphon first bends in the direction of the prey, and then rapidly extends as the septum contracts, sucking in its prey. It is a carnivorous clam.

Challenger was still in the Guinea Current, moving south and east around the bulge of Africa. She was now sailing across the abyssal plain of the Sierra

Leone Basin, floored with ooze and sandy mud, less than 200 miles from the African coast. This began to worry the scientifics, who had taken out medical insurance for the voyage, and now found a clause rendering it void if they landed on the west coast of Africa. Nares was following Admiralty instructions, however, and continued to sail on the starboard tack, close-hauled to the prevailing south-southwest wind. By 19 August the ship was beginning to feel the southeast trade winds and moved into the South Equatorial Current, which would take it west across the Atlantic, more or less on the Equator, bound for the coast of Brazil.

Station 101: 19 August 1873

05°48′ N, 14°20′ W; Sierra Leone Basin, 2,500 fathoms

At this hinge point the sails were furled early in the morning, steam got up, and the trawl lowered into deep water. It was left to work for the rest of the day and not recovered until late afternoon; a single cast of the trawl or dredge always took up most of the day. This time, it brought up a mixed bag of animals. There were bristlemouths, which had often been caught before, and a snipe eel, a long, thin fish with long, thin diverging jaws that forages a few hundred fathoms deep. There were also some bright red prawns, including one of the 30 or so species of *Gennadas*, all of which look more or less the same except for their highly specialized male and female structures. Mating in these animals involves the transfer of a ball of sperm, the spermatophore, from male to female, which she then uses to fertilize her eggs (this is not an unusual practice—it occurs, for example, in animals as different as spiders and salamanders). The male structure is a complicated appendage, which prawn biologists call the petasma, borne on the first pair of swimming legs, at the beginning of the abdomen; the female structure, called the thelycum, consists of modified plates at the base of the thorax. During mating, a male and a newly molted female (her thelycum must be soft) swim belly to belly as the male uses his petasma to thrust a spermatophore into the thelycum of the female. Naturally, this arrangement will work only if the petasma fits the thelycum. If the thelycum should become modified in some local population, copulation will become inefficient, and as a result there will be strong sexual selection that favors males with an appropriately modified petasma. In this way, sexual competition gives rise to a horde of species which look

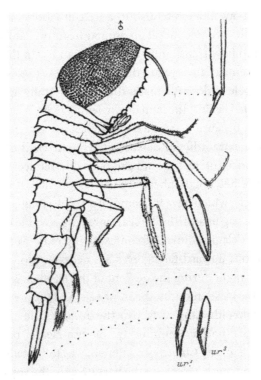

Figure 24 *Cystisoma* from Station 101
Source: T. R. R. Stebbing (1888), Report on Amphipoda. Zoology, Part 67, Plate 154. Bound in
Reports, vol. 29. Also in Narrative, vol. 1, Part 2 Figure 209.

almost identical except for structures that are used directly or indirectly in
copulation; a similar process of sexual selection drives species formation in
organisms as different as orchids and dung beetles.

A cloak of invisibility

Another shrimp-like animal caught by the trawl, far from being bright red,
was as transparent as it is possible for an animal to be. This was *Cystisoma*,
an amphipod, related to sand fleas and beach-hoppers and the out-of-place
animal that my grandson found in Green Park. It was very different, however, from the tiny animals that scurry away when you turn over a clump of
seaweed on a sandy beach. For one thing, it was much bigger, the size of a
child's hand. It was also so transparent—muscle, gut, and all—that it could

scarcely be made out in a white dish in a well-lit laboratory. Transparency often evolves in the sea to conceal an animal from predators, but there are normally limits to how transparent an animal can be. In the first place, the dark visual pigments of the eye are difficult to mask. *Cystisoma* gets round this by having a pale pink retina that is directed vertically upwards, to detect any prey silhouetted against the faint downwelling light. A carapace or cuticle is also a problem because its surface will reflect or scatter light. *Cystisoma* avoids this by the extraordinary device of having myriads of minute, evenly spaced protuberances on its legs and carapace that reduce reflection. It also has a layer of extremely small spherical particles, possibly symbiotic bacteria, covering the cuticle, which have the same effect. The result is an animal like a small clear plastic bag floating in the sea, almost impossible to discern.

Cystisoma is a pelagic animal, of course, living close to the limit of light where its enormous upward-facing eyes are necessary to detect its prey. It must have been caught during the descent of the trawl or when it was being recovered. Like the bristlemouths, the snipe eels, and the prawns, it lives deep in the sea, far from either the surface or the bottom. The existence of a numerous and diverse pelagic community was not clearly recognized at the time; the main task of the *Challenger*, after all, was to investigate the bottom of the deep sea, and one can sense in the narratives of the voyage a strong impulse to demonstrate that the animals found in the dredge or trawl really had been captured on the seabed. This seems reasonable, because the natural tendency of large objects in water is either to sink or to float, and any pelagic animal in deep water must contrive to do neither. Willemoes-Suhm was closer to the mark, though, when, at about this time, he remarked on the abundance and variety of organisms taken by a tow-net at 100 fathoms, and speculated that other animals might live still deeper. This topic was to become more urgent later in the voyage, when the tow-nets were set deeper.

Many animals manage to live in midwater because they are gelatinous, like jellyfish, ctenophores, siphonophores, and salps, which have nearly neutral buoyancy. Others belong to nongelatinous groups but have evolved some degree of jellification: one of the other animals captured at this station, for example, was a polychete annelid, about the size of a hot dog, whose body is covered by a transparent gelatinous sheath through which protrude long, jointed setae that help it to hang in the water column. *Cystisoma* is not gelatinous at all, and probably paddles along slowly using its swimming legs—living specimens brought into the laboratory can certainly do this. Its relatives are not gelatinous either, but many have evolved the extraordinary trick of using animals that are. Some latch onto a medusa and eat it as they are carried along. One finds a salp, like one of the units of a pyrosome, eats the animal, and then squats in the gelatinous tunic of its prey, riding around in a private submarine.

Gill slits

Another animal in the trawl, though, was certainly captured on the bottom. This was a puzzling worm (or, more precisely, the front half of a worm) called *Glandiceps*. It is a moderately large, unsegmented animal of simple construction equipped with a proboscis that it uses to burrow in soft sediment, which is why it was clear that it had been brought up from the seabed. It feeds by swallowing the sediment; but it can also draw in a current of water through its mouth, which passes through slits in the pharynx that act as gills and may also filter out edible particles, and is then expelled through pores in the body wall. This is not a very exciting way of life, but it is a very interesting one because it is how the most remote ancestors of chordates, the group that includes vertebrates like us, spent their time on the ancient sea floor, some 600 million years ago. Terrestrial vertebrates such as ourselves have long since given up straining mud, but our development still bears the stamp of this ancestral design: our faces have been modeled from gill slits and their supports, because evolution seldom innovates but rather modifies existing body parts. *Glandiceps* and its relatives are the hemichordates, a name that acknowledges their kinship with the chordates while keeping them at a respectful distance. They are not obviously attractive animals, and I doubt anybody has ever tried for a second time to eat one, so they are valued only by zoologists. They are most common in shallow seas, even in the intertidal, and for many years this specimen was the only hemichordate known from deep water. It no longer exists: it was sent to Germany after the voyage and was lost, probably destroyed by bombing, during the Second World War. Fortunately, the species was rediscovered 136 years later from even deeper water in the Romanche Trench, some 700 miles southwest of the *Challenger* station, by a Russian surveying expedition. This was only the front half, too, broken off when the worm was dragged from its burrow; nobody has yet seen the hind parts of *Glandiceps abyssicola*.

Station 106: 25 August 1873

01°47′ N, 24°26′ W; St Paul Fracture Zone, 1,850 fathoms

Challenger now turned west, out of the southern end of the Sierra Leone Basin, and sailed parallel to the St Paul Fracture Zone, where the sea floor is fissured and split into long trenches extending to either side of the Ridge. The regular trade winds and the strong current pushed the ship westward, just above the Equator. The scientifics and the officers must have been in a relaxed mood, for they were (in true British fashion) placing bets on what would be dredged up, payable (also in true British fashion) in drinks. I have no idea what sort of book they were running—a glass of sherry that there are no ophiuroids, two glasses that there is at least one glass sponge, that sort of thing. At all events, Wyville Thomson plunged heavily at this station, where he bet a bottle of champagne that there would be no umbellulid, and lost.

It was, in fact, an unusually fruitful haul. Besides the umbellulid, it included "a new and most hideous species of mudfish" (Campbell), which was another deepwater angler, and an unusual holothuroid, or sea cucumber. These are not normally the most winsome of creatures, just dark

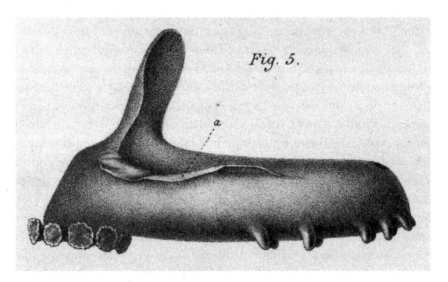

Figure 25 *Peniagone* from Station 106

Source: H. Theel (1882), Report on Holothuroidea. First, Part: the Elasipoda. Zoology, Part 13, Plate 10. Bound in Reports, vol. 4.

elongated bags lying on the seabed, but this one, *Peniagone*, was different. It is a translucent animal that hangs vertically just above the seabed, drifting along by using a sort of body flange as a sail to catch the weak current, and dipping down from time to time to grasp a sponge or one of its recumbent relatives. We now know that herds of these strange, ghostly creatures patrol the abyssal plain. There were also two stalked crinoids, which were always greeted with great enthusiasm by the naturalists, as a sort of

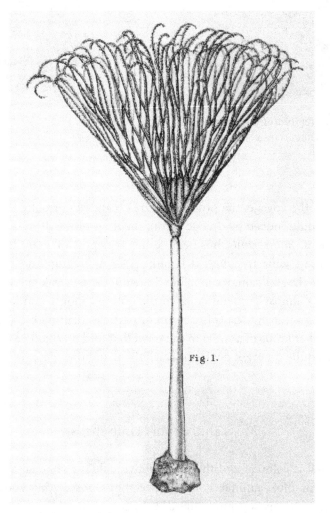

Figure 26 *Bugula* (Bryozoa) from Station 106

Source: G. Busk (1884), Report on Polyzoa. First, Part. Cheilostomata. Zoology, Part 30, Plate 8. Bound in Reports, vol. 10.

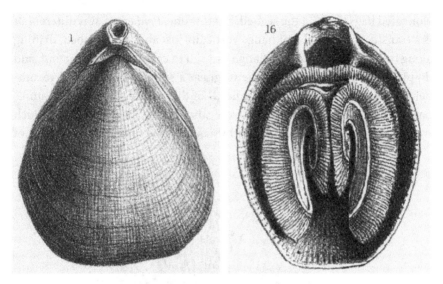

Figure 27 *Terebratulina* (Brachiopoda) from Station 106
Source: T. Davidson (1880), Report on Brachiopoda. Zoology, Part 1, Plate 1. Bound in
Reports, vol. 1.

emblem of the voyage, the proof that rare creatures normally found only
as fossils could indeed be dredged from the floor of the deep sea. Indeed,
one of them, *Bathycrinus*, has been found at depths of 4,000 fathoms or
more in the Pacific trenches. It is quite a small creature, no more than
about six inches tall, and its stem bends gently in any weak current so that
its crown is almost perpendicular to the bottom, mouth trailing, so that
its arms can intercept the thin stream of particles drifting by. There were
also two other kinds of animal in the trawl that feed on small particles, but
they use a quite different device to catch them and they deserve a separate
description.

Moss animals and lamp shells

One animal is colonial and the other individual. The colonial animal is a
bryozoan, or moss animal, consisting of thousands of tiny zooids, each
crouching in a stony cup, rather like a bushy coral. The other is a much larger
animal that looks superficially like a clam, enclosed by two bowl-shaped
shells that look a little like a Roman oil lamp. The two animals could scarcely

be more different in appearance, but they use the same device to collect food, a fringe of hollow ciliated tentacles mounted on a spiral or horseshoe-shaped base called a lophophore. The basic principle of the lophophore is simply stated—the particles in a current of water set in motion by cilia are trapped in mucus and swept into the mouth—but this may be misleading, because most particles are inedible, and an efficient food-collecting device would be able to reject them. I have struggled to understand how lophophores work, because they create very complex dynamic fields of moving water, and I may well have failed, but I think that they operate as inertial sorting machines, using a principle familiar to engineers. One familiar example would be winnowing: when chaff and grain are flung upward into a breeze, the grain falls out while the chaff is carried away by the wind. Another is the inability of large vehicles to turn tight corners at high speed: inertia forces them to continue in the direction they entered, so they leave the road. In a lophophore, the current that flows down the inner surface of the tentacles, impelled by cilia, meets a sort of lip or shelf above the mouth, so bends at a right angle before turning again and flowing across the underside of the shelf toward the mouth. Large particles fail to make the first bend and crash into the top of the shelf, and the cilia there sweep them away from the mouth and out via the bases of the tentacles. Small particles remain entrained by the flow, and the cilia on the lower surface of the shelf sweep them into the mouth. At least, that's how I think that lophophores work. Billions of these miniature machines daily comb the seas, as they have been doing for the last 400 or 500 million years.

Naming

In all, Murray lists 38 species from this station, of which 28 were new, including both of the crinoids, and had to be named. The naming process is as formal as a court, and likewise often leads to squabbles. Every newly discovered species is given a permanent name by the taxonomist who first publishes a formal description, using a specimen which from then on is the "type" with which any subsequent specimen thought to belong to the same species can be compared. The name consists of two parts, following the system established by Carl Linnaeus in the eighteenth century, the first giving the genus and the second the species. Grouping similar species into a common genus is a compromise between naming every species with a single unique word (unrealistic, because you would run out of words) or with a long string of words (clumsy and difficult to remember). The year of publication of the original description and the name of the person describing it are also often given in formal accounts. Thus, the common octopus of the Mediterranean is *Octopus vulgaris* Cuvier, 1797. The initial letter of the genus is always capitalized; the initial letter of the species is never capitalized; the two are written in italics, unless they appear in a sentence written in italics, in which case they are written in Roman. There are no exceptions to these rules, although the authority and year are usually omitted except in the most formal accounts. These simple rules have one major complication: the name ascribed to an animal often changes because the name originally given was preempted (the same name had been previously applied to another species), or because the opinion of taxonomists regarding the genus has since been overturned, or for any other bloody reason. Campbell's crinoid is a case in point.

It is quite normal to use the Latinized version of a personal name in order to honor someone connected with the discovery of a new species (the nearest equivalent in everyday life might be to have a local street named after you) and it was Wyville Thomson's habitual practice "to associate the names of those naval officers who have been chiefly concerned in carrying out the sounding, dredging, and trawling operations with the new species whose discovery is due to the patience and ability with which they have performed their task," so many of the officers of the *Challenger* found their names attached to one or another kind of sea creature. One of the crinoids found at this station, for example, was named *Hyocrinus bethellianus* by Wyville Thomson to honor G. R. Bethell, one of the lieutenants. It is a kind of immortality. The other crinoid is listed by Murray as *Bathycrinus campbellianus*, conferring a similar immortality on the aristocratic sublieutenant Lord George Campbell. But what's in a name? This one had been given originally by Carpenter in 1884 to the specimen found at Station 106. The complication was that Wyville Thomson had himself used the same specimen in 1876 to describe another species, *Bathycrinus aldrichianus*, captured later in the year at Station 146, far to the south, which he named for Pelham Aldrich, another of the lieutenants on the *Challenger*. Carpenter argued that Thomson's figure combined the crown of one specimen with the stem of another and that the woodcuts were unreliable, or poorly reproduced, and separated the Station 106 specimen as *B. campbellianus* from the Station 146 specimen as *B. aldrichianus*. Later taxonomists did not agree with him, however, and merged the two under a single name, which had to be *B. aldrichianus* because it was published first.

Thus it was that Campbell lost the contest of names to Aldrich. Immortality is not what it used to be.

Station 109: 28 August 1873

0°′55′ N, 29°22′ W; St Paul's Rocks, 104 fathoms

Two days later, half a dozen dark lumps appeared just ahead, the minute un-expected specks of St Paul's Rocks. These uninhabited and uninhabitable is-lands lie in the middle of the Atlantic Ocean, more or less on the equator, rearing up suddenly from abyssal depths. They are not volcanic; a sliver of mantle has been pushed through the ocean crust by tectonic forces and protrudes just above the surface of the sea. The entire archipelago would fit comfortably into a few city blocks, and its highest point would be overlooked by a six-story apartment building. It is naked rock, glazed with the dung of seabirds; there is no fresh water, nor any scrap of vegetation, not even a tuft of grass, and except for the highest ground the rock is swept continually by the Atlantic spray. It is, unsurprisingly, seldom visited; Darwin had touched there, 40 years before, and the next systematic survey would be a Cambridge University expedition a century later, in 1979.

Challenger approached from the west and moored in 100-fathom water, although less than 100 yards from the shore, by leading a hawser to a loop of line that Bethell had wound around a crag. This might seem to be appallingly dangerous—if the ship were to be blown on shore it would quickly disinte-grate, leaving any survivors marooned on waterless rocks that other ships took care to avoid. Nares counted, quite correctly, on the steady trade wind and the invariable current that flows westward at walking pace, although it must nevertheless have seemed strange to the crew to be moored to a rock in the middle of the ocean. They scrambled ashore, leaping onto a ledge as the ship's boat rose on a wave, to slither among the packed nests of noddies (like gulls or terns) and boobies (like gannets). Nares had issued strict orders that the birds were not to be harmed, but nobody took much notice, since the journals agree that the birds were at first so unsuspecting that they could be knocked down with a stick, which you would not know unless you had tried. By the second day they had become wilder, to the point where Moseley complained of the difficulty of shooting a few, for stuffing. Officers and men alike fished for black jack from a boat, or from the rocks, and took scores of the big, strong fish, although opinions differed about their edibility, from "rank bad eating" (Swire) to "the flavour is good" (Wyville Thomson). Many were taken by the small sharks that swarmed below before they could be landed, and once in a while the bait would be seized by a much larger shark that simply swam away irresistibly until the line was bent around a cleat for the break. Moseley (of course) fished for the black jack with a salmon rod, which he broke by giving too much butt, without any realistic prospect of being able to replace it.

Sally Lightfoot

Besides the birds, the sailors become acquainted with Sally Lightfoot, *Grapsus*, an amphibious crab with bright red legs and a yellow-spotted carapace, looking, at least among the more photogenic specimens, like an enameled clockwork toy. They swarmed in the crannies and crevices of the rock, feeding on flying fish dropped by the birds, and on the corpses of the birds themselves. They quickly identified the visitors as a new and valuable source of food, and clustered around the sailors at a few yards' distance, claws upraised, ready to spring forward if anything were discarded or to scuttle into the nearest crevice at the first sign of danger. Anyone sitting on the rocks by the water's edge to fish, and depositing the catch behind them, would turn to find crabs tearing at the gills of the fish, but retreating too rapidly to be caught themselves. Like all crabs, they used their claws, hinged limb tips toughened with chitin and powered by massive muscles, to break up large food items into fragments small enough to eat. Sally Lightfoot is a small crab about a forefinger in length, and its claws exert a force of no more than 10 newton (one newton is the force required to accelerate a mass of one kilogram one meter per second per second, say the force exerted by a large apple held on the palm of your hand). A large Atlantic shore crab might manage 100 newtons, enough to give a painful nip, and a lobster 250 newtons; the champion among crabs is the coconut crab, which can cut holes in coconuts using a force of about 1,800 newtons. For comparison, human grasp strength is only about 400 newtons for men and 300 newtons for women, despite our much larger size; the human hand has evolved for fine manipulation rather than crushing strength.

The animals living near to the shore were sampled by taking the dredge out in a boat out a quarter of a mile or so from the ship, dropping it overboard, and then recovering it with the donkey engine. This gave surprising results: in total, 16 species were recorded from a series of dredges in shallow water. This was fewer than expected; far fewer than had been found at comparable depths off islands like Madeira, or the Azores, or the coast of the mainland. St Paul's Rocks have, in fact, fewer species of fish than any other tropical island, and only a single endemic. Their situation is indisputably tropical, but in many respects the Rocks are more like an island in the temperate zone. The subtidal slopes, for example, are dominated by *Caulerpa*, a seaweed that not only has an interesting life cycle but also such vigorous vegetative spread, by stolons, that it can quickly cover large areas of the sea floor, to the dismay of anyone who disapproves of abundant organisms. (There is no doubt that that the manufacturers of artificial turf, should they ever turn their attention to seaweed, would take *Caulerpa* as their model.)

The remoteness of the Rocks explains the paucity of their fauna: they are a tiny target, many hundreds of miles distant from any source of immigrants. The few immigrants who survived the journey would be unlikely to find a mate, and if they did manage to breed their descendants would for long be so few that they would be easily extinguished by any slight shift in conditions, or simply by a run of bad luck. The scanty haul is easily explained, but it is the exception that proves the rule, a rule that had already been thoroughly documented by the voyage. Animals did indeed flourish at abyssal depths, far below the limit set by Forbes, but as a general rule their abundance and diversity decreased sharply with depth; whereas 100 species might be taken by a trawl in shallow water, barely 10, or fewer, would be recovered from a deep trawl, and perhaps not all of these really lived on the bottom. Maybe Forbes was right after all, but had simply underestimated the depth limit. The deep water of the Pacific, which *Challenger* had yet to explore, would no doubt settle the question.

Station 122: 10 September 1873

09°05′ S, 34°50′ W; Recife Plateau, 32 fathoms

Casting off the hawser, *Challenger* left the Rocks and slanted southwest across the Fernando de Noronho Abyssal Plain. She crossed the Equator for the first time on the voyage around midday on 30 August, but Nares forbade the traditional initiation ceremonies (such as being shaved with a rusty barrel-hoop) and made an issue of wine instead. There were no complaints. The sky changed, with the North Star fading, to be replaced by the Southern Cross, a ponderable change for sailors who steered by the stars. Several soundings were taken in deep water, running through the now familiar sequence of pteropod ooze to globigerina ooze to red clay, without any dredging being attempted, perhaps in haste to make the next landfall at Fernando de Noronha, which was sighted early in the morning of 1 September.

This was a more conventional tropical island, or rather archipelago of volcanic peaks, lying off the coast of Brazil. It had been heavily forested when Darwin visited in the *Beagle*, but it now served as a penal colony, and much of the forest had been cut down, perhaps to deny a refuge to escaping prisoners, or deprive them of timber to build boats. Enough remained for the naturalists to explore, and they looked forward keenly to collecting the fauna of this little-known, secluded place. Moseley even accompanied Nares to interview the Governor of the island, although he was plunged up to his neck as the boat landed in heavy surf, and must have been at least slightly damp. At this time, the island was administered by the Ministry of War, and the Governor, a military appointee, was puzzled. Why had an English man-of-war suddenly arrived off his island, and why had no salute been made? The concept of a man-of-war from which almost all guns had been removed, making the usual salute (15 guns, at a guess) impracticable, and was instead fitted out for scientific exploration, proved difficult to explain, especially as the Governor did not speak English and Nares had no Portuguese. The interpreters did their best, though, and over coffee and cake it was agreed that the naturalists could explore the island and would even be provided with guides. The ship's party returned in high spirits to make their preparations.

Early the next morning, word came that the Governor had changed his mind, and would not now allow any exploring or collecting. I am not surprised. There was (unknown to Nares) a history of resistance and rebellion among the prisoners, and the Governor must have reflected, or been advised, that should these strange activities give any opportunity for escape he would have to answer to the Minister, whereas the foreigners would have long since departed. In fact, shore parties had already left, and Nares went ashore to petition the Governor, but to no effect, and the parties were recalled. Half an hour later the ship, with a disappointed crew, set sail for the mainland.

On the way out, Nares could not resist landing on a couple of the uninhabited outlying islands—who was to know?—which Moseley searched without finding anything of great interest. Having thumbed his nose, he set all sail

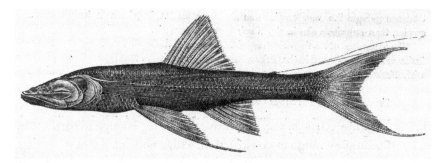

Figure 28 *Bathypterois* from Station 133
Source: A. Gunther (1887), Report on Deep-Sea Fishes. Zoology, Part 52, Plate 47. Bound in
Reports, vol. 27. Also in Narrative, vol. 1, Part 1 Figure 88.

and made for the coast, which was sighted two days later as a dark bar of
forest on the horizon. Their route took the ship over deep water in the unnamed basin off the easternmost bulge of South America, but no sampling
was attempted, perhaps because it had now come on to rain, and the prospect
of sifting through mud in heavy rain on the unprotected dredging platform
could not have been very attractive. By 9 September the ship was running
south parallel with the coast, in relatively shallow water near the edge of the
continental shelf, but the wind had by now turned foul, and any further progress had to be mainly by steam. This at least gave an opportunity to use the
trawl, however, and, in strong contrast to the meager fauna at St Paul's Rocks,
a series of hauls over the next couple of days brought up more than 200 species, including stalked crinoids, a very curious red worm that seems to have
been lost, the first octopus, a tripod fish, and a huge barrel sponge.

Tripod

The tripod fish, *Bathypterois*, is a quite common deepwater fish about the size
of a herring. The lower rays of the pelvic and caudal fins have become enor-
mously lengthened—they can be longer than the body—to form a tripod
that supports the fish on soft sediment. The pectoral fin rays have also become lengthened, and extend in front of the fish (which is practically blind)
as feelers to sense moving prey. It crouches on the bottom, in the darkness,
until a copepod or minute shrimp swims by and is detected by its feelers and
sucked in. Like many deepwater fish it is hermaphroditic, probably because

this increases the chance of finding a mating partner when neighbors are scarce and rarely encountered.

Fine filtration

The barrel sponge, *Geodia*, was impressive in size, big enough to bath a baby in, but it is simple enough in construction, with a body plan that dates back to the origin of animals. Six hundred million years ago there were no fish in the sea, nor any animal of any kind, and a surveying party of extraterrestrial naturalists would have found nothing to dredge except mud. They might not even have noticed an inconspicuous group of single-celled organisms that propelled themselves through the water by a long flagellum beating inside a sort of cup attached to the surface of the cell. These are choanoflagellates. The cup consists of closely spaced rods, so closely spaced that when the current of water produced by the flagellum passes between them, any bacteria are screened out and can be absorbed by the cell. Many choanoflagellates tend to stick together after cell division, forming little colonies. Once upon a time, one lineage formed colonies in which choanocytes, the individual flagellated cells, lined the inside of a hollow chamber, with pores on its outer surface to let the water in and a common exit to let it out again. This fairly simple modification is more efficient because a group of cells arranged like this can generate higher pressure than isolated cells. What it creates, however, is no longer a choanoflagellate, but rather a small multicellular creature with all the essential features of a sponge—perhaps the earliest animal, in fact. This is not merely imagination; a minute fossil of this kind has recently been found in Chinese deposits nearly 600 million years old, in which, quite exceptionally, details of cellular structure have been preserved. Modern sponges are the enlarged descendants of these earliest animals that have retained the body plan of a hollow perforated vase through which a stream of water is forced. They are, in effect, animated water filters, and very efficient ones too, capable of clearing hundreds of times their own volume of seawater per hour.

The sponge way of life has one serious drawback, though: they are quite unable to move, lacking both muscle and nerve, and can summon up no more than a slow-motion shudder at best. None have evolved a protective covering like the stony armor of corals and bryozoans, or the shells of clams and brachiopods. Consequently, they were very vulnerable to animals with jaws, which soon evolved (well, within 50 million years or so). Their riposte was spicules: minute shards of calcium carbonate, or silica, that permeate the sponge body, imparting a certain firmness, but also making eating a sponge like eating sandpaper. Just as bad, though, passing a stream of seawater though a rather loosely organized body gives bacteria a good opportunity to take up residence rather than being eaten, and sponges are in fact clogged with a whole zoo of bacteria, like a massive lifelong infection. They protect themselves (perhaps) with a whole pharmacopeia of biologically active chemicals—antibiotics, antivirals, antioxidants, antitumor compounds, immunosuppressors, and more, all potentially of interest to pharmaceutical companies. The technology is brutally simply: grind up a sponge and see if you can find a useful compound in the mash. How any of them contribute to sponge well-being is rarely known and rarely asked, an interesting comment on the current state of our understanding of biochemistry; why an animal without an immune system should produce immunosuppressors, for example, is far from clear. It is often not known even if these compounds are produced by the sponge itself, rather than by one of its innumerable bacterial lodgers. Sponges have survived half a billion years and continue to thrive, despite being unable to bite or run away, because nasty chemicals and sharp spicules combine to make them unpleasant and unrewarding to eat. Some sea turtles browse on sponges, and a few specialist fishes, but so far as I know they are the only abundant and diverse group of animals that does not feature in the cuisine of any human society.

The ship limped on in a failing wind, sailing when possible but mostly under steam, toward Salvador (always called Bahia by the crew). She came in sight of the harbor with two buckets of coal left, which lasted only 10 minutes and left her becalmed most of the day, surrounded by a cloud of tropical butterflies. At last, an evening breeze filled the sails and allowed her to creep in and anchor half a mile from the shore.

Challenger remained at anchor for the next two weeks, the first prolonged stay of the voyage in a tropical port. The crew ran ashore and were impressed by the town, except for the high prices and the fearful stench of the narrow streets. Two did not return, but having bad characters were not eagerly sought for. Willemoes-Suhm bought four stuffed hummingbirds for his mother to wear in her hat. Moseley bought a sloth and kept it in his cabin for a while, but it would not feed, and was eventually added to the collections. Wyville Thomson, Moseley, and the others went on excursions by rail, horseback, and coastal steamer. A game of cricket was played against the Bahia Cricket Club; the British influence was strong at this time, and introduced both cricket and soccer to Brazil, with widely divergent results. Wyville Thomson had not forgotten his scheme to equip the scientifics with a uniform, and one day "our worthy but very vain old professor" (Swire) appeared on deck in the shell jacket of a volunteer colonel of artillery, with a pair of yellow holland trousers. He was "greeted by the suppressed tittering of the assembled officers."

An American frigate arrived on 16 September and set in train a social whirl. There was to be a ball on shore, given by the cricketers, and a play on board the frigate to entertain its crew, who were not allowed on shore because so many had deserted at Rio. Neither event took place. A week later, one of the seamen returning from shore leave fell sick with yellow fever, and Nares had to make a very quick decision. All shore leave was canceled; the men remaining on shore were brought back to the ship; the patient was returned to a hospital on shore, where he died a few days later; and before the light failed the ship had weighed anchor and moved out of the bay. If he had hesitated, the ship would have been held for weeks in quarantine, completely dependent on the shore, and the voyage, with its carefully planned schedule, thrown into disarray. It was a gamble, of course: if the crew came down with yellow fever the ship would become a floating mortuary, with, in the worst case, too few sailors to man the sails or stoke the boilers.

Nares set a course south toward colder seas, abandoning a planned visit to Trinidad, and put out the fires in the boilers to cool the ship (all sound procedures: yellow fever is spread by a tropical mosquito). The second case of fever appeared a few days later. The patient, a Marine, was put in his hammock on the upper deck, partly for the cooling airs but partly also, no doubt, to isolate him from his shipmates. He recovered. Nares himself then went down with a fever of some sort, but threw it off; and after that there were no more cases. The gamble had paid off.

10

Fifth Leg

Across the South Atlantic, September–December 1873

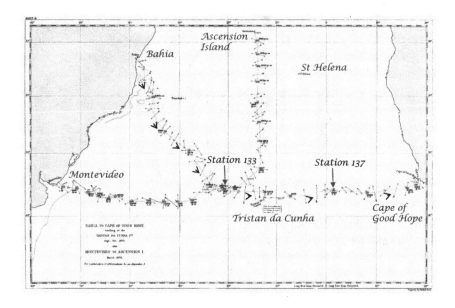

Challenger raced southward along the western edge of the Brazil Basin, skirting the shallow banks and shoals that line the coast here. She literally raced some of the ships she met, although she was too lightly rigged to be very fast; she managed to outpace *Prince Arthur*, but this ship was 52 days out from Liverpool and no doubt foul. Nares was in too much of a hurry to reach colder regions for much dredging to be done, and what was attempted was unsuccessful. On 30 September, approaching the chain of Vitoria Trindade Seamounts, the dredge rope carried away in 2,150 fathoms and all the gear was lost. This was disappointing, but Willemoes-Suhm had caught something interesting in a surface tow: a small insect.

Full Fathom 5000. Graham Bell, Oxford University Press. © Graham Bell 2022.
DOI: 10.1093/oso/9780197541579.003.0011

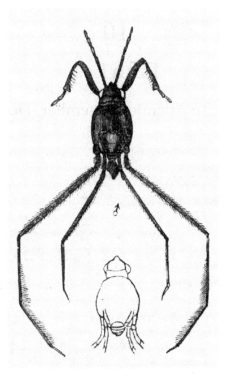

Figure 29 *Halobates* from surface tow.
Source: Narrative, vol. 1, Part 2 Figure 179.

Sea skaters

A small insect would not be anything remarkable on land: a few passes with a sweep-net through tall grass will supply hundreds. Insects have never taken to the sea, however. There are quite a few kinds of insect able to potter around on beaches or in mangrove swamps, but only one that is truly marine and lives all its life on the open ocean, and this is the animal that had turned up in Willemoes-Suhm's tow-net.

Any pond or patch of still water will have a ballet of pond skaters, dancing around on long thin legs but also, when they come across their prey, driving a hardened tube into it, liquefying its flesh, and sucking in their meal. Some of them, called *Halobates*, have gone to sea, where their habits are similar. They have short front legs for finding and grasping prey (or females), middle legs for sculling (they can move, jerkily, at up to one meter per second), and hind legs for steering. Legs and body are covered with water-repellent hairs that keep them afloat as they search for anything trapped in the surface film. They lay their eggs on feathers or pieces of wood, or nowadays on plastic waste or tar lumps, and live all their lives at sea. They live in Flatland, a continuous and, from their point of view, infinite two-dimensional landscape from which they can never escape. They can jump a little, perhaps to escape some threatening fish, but fall back immediately; if submerged accidentally by a wave they pop up again like a bubble. Most of the time, they skim over the heaving surface of the sea, sensing opportunity and danger by reading the ripples with their sensitive front legs. Their enemy is the blue noddy, a minute seabird scarcely two ounces in weight, which sometimes feeds almost exclusively on sea skaters, which have half a second to realize that they have just been scooped up by something from another dimension that, inexplicably, made no ripples at all.

Three days later, just north of the Rio Grande Rise, the trawl was brought up from 2,350 fathoms, but just as it came to the surface a swivel broke and, full of mud and, no doubt, animals, it sank slowly back into the depths as the naturalists stared blankly and Campbell, who had been responsible all day for working it, cursed. They tried again after another three days, this time bringing the trawl up successfully from 2,275 fathoms, only to find that it was empty apart from two pumice pebbles and a few fragments of whalebone. Fourth time lucky.

Station 133: 11 October 1873

35°41′ S, 20°55′ W; mid-Atlantic Ocean, 1,900 fathoms

At this point, *Challenger*, heading southeast, was again sailing above the fault-scarred terrain on the approach to the Mid-Atlantic Ridge. For the first time in four or five days the sails were furled in the morning, the trawl put over and safely recovered seven hours later, full of animals but with nothing really new. It was interesting that there should be nothing really new.

A cosmopolitan cucumber

For example, there was a holothurian, or sea cucumber, called *Psychropotes*. Holothurians are echinoderms, like starfish or sea urchins, but are quite different in appearance, resembling large leathery sausages lying on the seabed. They are mostly deposit feeders of the simplest kind: the mouth is surrounded by a ring of tentacles—modified tube feet—that shovel in sediment from which the organic matter is digested and the rest, which is most of it, expelled. *Psychropotes* is easily recognized because it bears a large finlike structure on its back, looking something like the tail of an airplane; no one seems to know what it is used for. The naturalists had seen it before and were to see it again on several occasions. It can be found, in fact, living at abyssal depths in all the oceans of the world, a cosmopolitan deep-sea animal. The solitary coral *Fungiacyathus* came up in this haul too, and is likewise found throughout the world in the deep sea. The blind lobster *Willemoesia* turned up again, and would be found once more faraway in the South Pacific. A hermit crab whose name, *Parapagurus abyssorum*, reflects its home in deep water, was collected repeatedly in the Atlantic and Pacific. The scale worm *Laetmonice* that was caught here had been encountered previously off the Azores and is also found close to Antarctic ice and in the tropical Pacific. Willemoes-Suhm remarked that the South Atlantic deep-sea fauna was similar to that found in the North Atlantic. Moseley went further: "The universal distribution of deep-sea forms becomes more and more apparent."

These are interesting claims because they emphatically do not apply to terrestrial animals. The fauna of North America is quite different from that of South America; and in an east-west direction North America and Europe, or South America and Africa, are equally distinct. Given a random mammal, lizard, or freshwater fish, its continent of origin can in most cases be confidently identified. These distinctive faunas have evolved because dispersal between continents is very infrequent, or unknown, so that the same way of life can evolve independently in different places. Hummingbirds feed on nectar in North America and sunbirds do the same in Africa; anteaters dig up ants in South America and aardvarks dig up termites in Africa; electric fish live in the rivers and swamps of both South America and Africa. These similarities do not imply relatedness, but rather the repeated evolution of similar bodies to fit similar ways of life in isolated communities. Dispersal is much easier for marine animals, especially those with pelagic larvae that may be carried hundreds or thousands of kilometers by ocean currents, and yet there is a profound difference between a coral reef and the coast of Labrador, whose communities may have no species in common. The difference in this case does not arise from the difficulty of dispersal, but rather from adaptation to different conditions of life. In the deep sea, however, neither barriers to dispersal nor differences in conditions are nearly as pronounced. The abyssal plain is an essentially continuous worldwide habitat, divided only by rift zones that are too narrow to offer much impediment to dispersal. The physical conditions of life are everywhere similar: it is cold, dark, unproductive, and floored with soft sediment. The conclusion that the naturalists on board the *Challenger* were feeling their way toward was that these features of the deep sea imply that the geography of the animals that live there is unique: instead of the sharply distinct regional faunas of land or coast, there is instead a universal fauna of the abyss. We shall see later how well this theory held up as they explored the depths of the Southern Ocean and the Pacific.

Figure 30 *Ipnops* from Station 133

Source: A. Gunther (1887), Report on Deep-Sea Fishes. Zoology, Part 52, Plate 49. Bound in Reports, vol. 27.

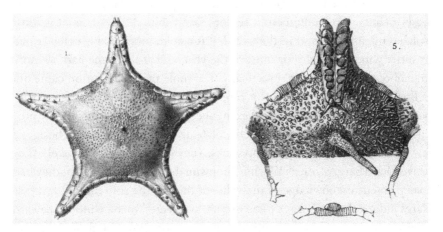

Figure 31 *Porcellanaster* (and its snorkel) from Station 137

Source: T Davidson (1889), Report on Asteroidea. Zoology, Part 51, Plates 20 and 21. Bound in Reports, vol. 30.

Neither blind nor sighted

There was also a long, dark fish flopping around in the trawl that had never been seen before. Disregarding the normal categories, it was not blind, nor was it sighted. Named as *Ipnops*, it bore on its head a pair of flat plates (first described by Moseley) whose function was long debated. They are now known to be eyes that have been modified almost out of recognition, lacking lens and iris and cornea, but still connected to an optic nerve and directed resolutely and immovably upward. From being like a camera they have become more like a scintillation counter, able only to detect the faint luminous spark of a passing shrimp, as the fish lies on the seabed, staring lidlessly upward on the seabed until this faint spark passes and then lunging upward to seize its prey.

It now began to get much colder. The first albatrosses were seen, and cold-weather clothes were issued. It began to rain and blow, forcing the ship to reef topsails, but 13 October dawned fine and clear, although very cold to men accustomed to the tropics, and early in the morning of the following day Tristan da Cunha was sighted, a dark smear at first, but as they approached more closely, revealing itself as a huge volcanic cone 8,000 feet in height, with permanent snow on its peak. The ship laid to for the night, and steamed into Edinburgh Bay at first light, anchoring half a mile off the settlement.

The island group lies in the middle of the South Atlantic, practically on a line drawn between Cape Horn and the Cape of Good Hope, and since the sea elephants were hunted out it seldom saw a ship. Besides being the most isolated inhabited island in the world, it is a cold, inhospitable place lashed by bitter winds and rain for most of the year, and at that time had no more than about 70 inhabitants. Practically the whole male population came out in boats, taking advantage of this rare visit from the outside world to barter potatoes, eggs, fowl, and pigs in exchange for supplies they could not produce for themselves: flour, sugar, chocolate, tea, wine, and tobacco, besides a few household items such as lamp wicks. They were no ignorant yokels: they drove a hard bargain, or, as Wyville Thomson delicately expresses it, they had a very practical knowledge of the value of things. The ship did not tarry, as Nares judged the anchorage to be unsafe, and feared that a wind to seaward might drive her on shore. Land parties were not allowed out of sight, and with the weather deteriorating, the ship weighed anchor and steamed off the same evening.

A tale of two brothers

Before they left, the crew heard a strange story. When the *Beacon Light*, out of St. John's, Newfoundland, took fire and sank in the South Atlantic in late 1870, her crew managed to reach Tristan in an open boat. They were later taken off by a passing ship and landed at Aden, from where one of them, Gustav Stoltenhoff, traveled back to his elder brother, Frederick, in Aix-la-Chapelle. Gustav persuaded Frederick to return with him to Tristan to make their fortune hunting seal, so in August 1871 they took a steamer to St. Helena, where they found an American whaler to take them the rest of the way. The captain of the whaler was much less optimistic than Gustav about how warm a welcome they could expect from the islanders, and suggested instead that they try Inaccessible, a small uninhabited island about 30 miles from Tristan, where there was sure to be plenty of seals and sea elephants. They agreed, and were landed on Inaccessible by ship's boat, with enough supplies to last them for the season, on 27 November 1871. The list of supplies (like all lists) is interesting:

An elderly whale boat (bought at St. Helena)
Two blankets each, boots, and the clothes they had traveled with
Lantern, oil, and matches
Wheelbarrow
Simple tools (spades, shovel, pickaxes, knives, saw, hammer and nails, chisels)
Door and glass for a window
Cooking utensils (kettle, frying pan, saucepans, and cutlery)
Empty barrels for oil
Short Enfield, muzzle-loading, and an old fowling piece, with powder and bullets
Flour (200 pounds)
Rice (200 pounds)
Biscuit (100 pounds)
Sugar (30 pounds)
Coffee and tea
Salt (one barrel)
Tobacco (eight pounds)
A few bottles of wine, gin, and vinegar
Medicine (Epsom salts)
Seed potatoes and vegetable seeds

As it turned out, these supplies, with very little replenishment, were to last them, not for a single season, but for the next two years.

A few days later, boats from Tristan arrived for the sealing season, having heard of them from the captain of the whaler. The visitors were friendly, and helped them move their supplies to a beach on the north side of the island, a better site, where they set to work building a hut and clearing and planting a vegetable plot. The weather was fine, there was an abundant supply of wood and water close at hand, and the feral goats and pigs on the inland hill could be shot for meat. The seals, however, were scarce and difficult, and the sea elephants had departed. In January the boat was damaged in the surf and could afterward only be used by constant bailing, until they cut it up to make a single ungainly boat. In April the tussock grass on the cliff that hemmed in the beach caught fire—they were clearing ground nearby by burning, and the sparks carried. This was a disaster, because they used the tussock to clamber up the cliff, and the hill was now inaccessible from the beach. They could still row round the point in their abbreviated boat, but in June it was smashed by a storm. They began to starve as their supplies ran out, and by August were mere skeletons, living on penguin eggs and thrushes. Time passed slowly as they sat in their hut, lit by a single lantern, while the wind raised the surf and the rain poured down. They were saved by the arrival of a French vessel, which bartered eggs for biscuit and tobacco, which sustained them until a boat arrived from Tristan with some salt pork, but little else, and the men were far from friendly, even stealing a few small items from the brothers. By November their supplies were exhausted, and in desperation they swam around the point and climbed the hill to shoot pig and goats. While they were there an American schooner appeared, and they got a few pounds of flour and molasses in exchange for half a dozen seal skins. All of the goats, though, were shot by parties from Tristan. By March they had run out of supplies again, even tobacco, and were forced to swim round the point again and afterward live on undiluted pig. It was only when the penguins began to lay at the end of the winter that they were able to change their diet to eggs fried in pig fat. By October 1873 they had no food left, their tools and guns were broken or lost, and only the Epsom salts and vinegar remained untouched. It was at this point that the *Challenger* appeared, after two years of hardship and solitude.

This, in brief, was the experience of the Stoltenhoff brothers, as narrated by Frederick to Paymaster Richards. Why the brothers did not take passage in one of the earlier ships, and whether the islanders from Tristan acted through malice or neglect, are difficult questions, but there is no doubt that their story is substantially true. The *Challenger* carried them to Cape Town, where Richards wrote an account of their super-Crusoe solitude for the *Cape Monthly Magazine*. Here they parted company, with Frederick staying on as an office clerk while Gustav returned home; so far as I know, they never met again.

Before leaving the group, *Challenger* called at the third island, Nightingale, to set up a surveying station on its rocky peak. The slopes leading up to the peak were practicable, and apparently covered in grass that, from a distance, offered an easy climb. Its appearance was misleading, however. In the first place, the tussock was taller than a man, growing in large clumps of coarse reed-like grass, almost woody at the base. Once you entered the tussock it was no longer possible to see your destination and you soon become disoriented. Second, the area is filled with nesting penguins. At first you step gingerly to avoid the birds, but they are so thick that it is impossible to make progress, so you push them out of the way, which they resent and peck at your ankles with their strong sharp beaks. The soil is black and slimy, and the air stinking and stifling, and the noise of 10,000 birds deafening, so after a while you blunder along as fast as you can, crushing eggs and young and belting the adults out of the way with your stick, until at last you stagger out of the rookery onto the seashore, lost and sore, with your destination as far away as ever. The surveyors returned defeated; no station was ever set up.

Station 137: 23 October 1873

35°59′ S, 01°14′ W; Cape Basin, 2,550 fathoms

With the brothers Stoltenhoff on board, the ship jogged north to pass close by Tristan, dredging on the way and getting hundreds of specimens (including more stalked crinoids) from depths between 100 and 1,000 fathoms. Wyville Thomson comments that the fauna would be difficult to distinguish from a similar haul from the coastal waters of the Mediterranean, or even (according to Willemoes-Suhm) England. They finally desisted on the evening of 18 October and set a course eastward to the Cape, pushed along swiftly by a fresh westerly breeze. They attempted to dredge two days later, but the dredge fouled something on the bottom and came up empty. A gale blew up in the evening and continued all the next day, with the ship making 10 knots under double-reefed topsails.

Snorkel starfish

The gale eventually subsided and the dredge was put over again on 23 October, in 2,550 fathoms, as they were entering the deep water of the Cape Basin, passing just south of Walvis Ridge. There were only a few animals in the ooze when it was recovered. One was a curious little starfish with broad-based pointed arms that give it a webbed appearance. This was *Porcellanaster*, one of a group that lives only in deep water. It seems to spend most of its life buried in the mud (and therefore not often captured by a trawl) and has a snorkel-like ventilating tube on its upper surface that extends into the overlying water. It seems to feed by simply packing the ooze into its stomach and digesting what it can before expelling the rest by the way it came (it lacks both intestine and anus) and shuffling off to a new spot. Mud is not a rewarding meal, and the stomach contents may weigh more than the animal itself. Most starfish are carnivores, of course, and so it is tempting to infer that deposit-feeding is an adaptation to the scarcity of prey in the deep sea. The clam *Rhinoclama* that was captured a couple of months earlier shows the opposite pattern, however, taking up a predatory mode of life in the deep sea although most bivalves are suspension-feeders (and another carnivorous deep-sea clam, *Policordia*, had been found just before Tristan, at Station 133). Any feature of a deep-sea animal might be a specific adaptation to deep-sea conditions; but the bare fact does not warrant the conclusion.

The wind was fair for Cape Town, but began to freshen until, with all sails set, the ship was making 14 knots. Swire remarks that it was not clear that the watch had understood the captain's instructions, but this seems a rather odd comment. There had been a few other incidents recently hinting that the cheerful saga recounted in most of the published journals might not be the whole story. When the coal ran out while steaming into Salvador, for example, Nares, it is said, delayed setting sail because he was unwilling to wake the crew from their siesta, which does not seem very Navy. Landing on Nightingale, "The bluejackets who landed with us . . . behaved with their usual freedom; for their officers they did not seem to care a fig." This might be no more than Swire being grumpy, but the continual trickle of desertions whenever the ship moored in a mainland port suggests that the *Challenger* was not entirely a happy ship. On this occasion the wind moderated on the following day, but there had

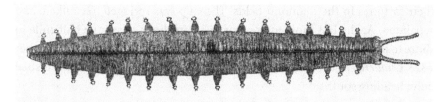

Figure 32 *Peripatus*, from Cape Town
Source: Narrative, vol. 1, Part 1 Figure 113.

been some damage to the main yard, and Maclear went around looking solemn.

Table Mountain was sighted on 28 October, and the ship anchored in Simon's Bay soon after. She was placed in quarantine because of the case of yellow fever at Salvador, but only for two days, after which the crew were free to run on shore. They had a long run, staying at anchor for the next seven weeks. Murray spent the time boxing up the specimens that had been collected so far and sending them off to Edinburgh. Moseley spent the time traveling around and shooting things, but also joined Willemoes-Suhm in searching for an extraordinary animal, an onychophoran called *Peripatus*. It had first been described, as an unusual kind of slug, only about 50 years previously, and indeed it looks like a rather flabby uninteresting slug a few inches in length, but it has antennae and a long row of stubby legs on each side of its body, a little like one of those fairground animals conjured out of a set of sausage-shaped balloons. Its main claim to fame is that it is the sister group of arthropods. Long ago, the common ancestor of all insects, crabs, shrimps, spiders, mites, and millipedes looked something like this, before those stubby legs evolved into stiff jointed limbs encased in cylinders of chitin. They are not easy to find; the usual short-cut of offering money to local boys did not work this time, and the two naturalists searched long and hard before finding specimens in decayed willow logs and under an old cart-wheel in a farmyard.

The officers amused themselves with cricket and dancing, ending in a grand ball, perhaps the only one ever held with dredging as a theme, at which the officers enjoyed themselves and the ship's band became so drunk they could no longer play or even hold their instruments. About a dozen men deserted, some preferring to spend a month in jail before taking passage in another ship rather than remaining in *Challenger*, others going off to seek their fortunes in the diamond fields. The ship was rerigged, recaulked, and repainted. At the last moment, one of the ship's boys swam the half-mile to shore to meet a woman, but was detected and brought back. On 17 December 1873 *Challenger* steamed out of Simon's Bay with a somewhat diminished crew, heading south.

11

Sixth Leg

The Southern Ocean, December 1873–April 1874

Challenger now left the familiar waters of the Atlantic and set out in fine weather into the wilder and less-explored region that circles the frozen continent of Antarctica. Her course was south, then east, toward the barren uninhabitable islands of the Southern Ocean: Marion, the Crozets, Kerguelen, and Heard, lonely places where only the whalers and sealers normally went. Heading west would have taken her through the narrow gap between Cape Horn and the Antarctic Peninsula, where the southern seas pile up, and she would have had to force her way against the prevailing westerlies, a famously long and perilous passage for a sailing ship, even one equipped with a small steam engine.

Sea serpent

The first animal seen as the ship steamed out of False Bay was a sea serpent, with its head and neck sticking two feet above the surface. Sightings of sea serpents were not uncommon at the time, perhaps because it was often not practicable to overhaul them quickly with a sailing ship. The frigate HMS *Daedalus* had seen something similar in the Atlantic between St. Helena and the Cape some 30 years previously. This one turned out to be an ocean sunfish, *Mola mola*, an extraordinary animal that looks like a child's first attempt to draw a fish. Its body is more or less circular, flattened side to side, with no real tail but two vertical flukes toward the rear (the modified dorsal and anal fins), the upper sticking out two feet or more above the surface. It is enormous enough—large specimens weigh as much as a small car—to make a very satisfactory monster.

Full Fathom 5000. Graham Bell, Oxford University Press. © Graham Bell 2022.
DOI: 10.1093/oso/9780197541579.003.0012

Station 143: 19 December 1873

36°48′ S, 19°24′ E; Agulhas Bank, 1,900 fathoms

Cucumbers with legs

A light breeze from the southeast soon carried the ship out of sight of land across the shoals of the Agulhas Bank before finding deeper water to the south. Dredging in shallow water on the bank had produced enough specimens to keep the naturalists employed for months, but the haul became leaner as the ship moved into deeper water and the animals began to change. Only three species came up from 1,900 fathoms at the southern edge of the bank, but one of them is worth a second look. It is a small translucent animal with tentacles and a row of stubby legs on either side of the body, looking, in fact, for all the world like an onychophoran of the sort that Moseley and Willemoes-Suhm found on the mainland, but it is nothing of the sort; emphatically not. It is, in fact, another holothurian, rather pejoratively called

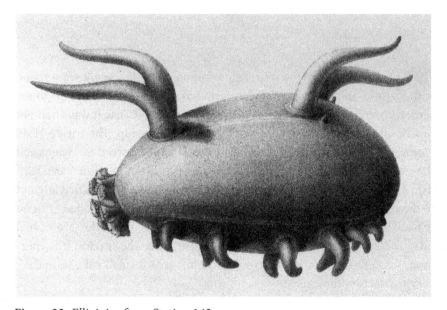

Figure 33 *Ellipinion* from Station 143

Source: H. Théel (1882), Report on Holothuroidea. First, Part. The Elasipoda. Zoology, Part 13, Plate 5. Bound in Reports, vol. 4.

a sea pig, *Ellipinion*, belonging to the same deep-sea family as the drifting ghosts found in August. Instead of drifting just above the seabed, however, it stumbles across the ooze using tube feet fantastically modified as inflatable legs. I have no idea how it flexes them so as to propel itself, but it does, engulfing the mud as it goes. Only the five tentacles surrounding its mouth betray the invariable pentaradial symmetry of an echinoderm. Oddly enough, onychophorans appear in the fossil record, very early indeed, as marine animals, before disappearing from the record for 500 million years, until they were discovered still living in tropical leaf-mold. Sea pigs give us some idea of how they would have lived on the seabed in the hectic time, half a billion years ago, when animal body plans were being tried out for the first time.

The weather was at first unusually fine, until shortly after the anniversary of the voyage, on 21 December, when the ship moved into the path of the westerly winds that sweep around the Southern Ocean. It became colder, and, anticipating the weather to come, warm clothing was issued: thick pea jacket, large, worsted comforter, knitted jersey, thicker drawers, sou'wester, and a pair of mitts. Stormy weather began to set in, and by Christmas Day it was freezing on deck, 20 degrees colder than just a few days before. Christmas dinner for the crew was salt pork and plum duff, washed down with Madeira. Officers and scientifics assembled in the wardroom for drinks at 6:00 p.m. I don't know what they ate, but I'm sure they had guarded it more carefully than last year.

Marion and Prince Edward islands were sighted while dinner was being served, and parties landed on Marion early the following morning, tugging the boats over the thick kelp and slippery rocks of the shore. These southern islands are all desolate places; Marion is no more distant from the Equator than Paris, but is blasted into tundra by the endless chill wind and rain. Moseley was with the first party, as usual, which immediately found a sea elephant and killed it with a rock. They struggled through the boggy ground, sweating in their heavy clothes because the day was unusually fine, killing the albatrosses whose nests were strewn thickly on the grassy slopes. The sailors killed them for their breast down, their wing bones (to make pipe stems), and their webbed feet (to make tobacco pouches); the naturalists killed them for science; Willemoes-Suhm expected that they would soon to go the way of the dodo. Meanwhile, the ship dredged the shallow channel between Marion and Prince Edward, where the volcanic sand was carpeted with bryozoans. The shore parties returned in the late afternoon and the ship made for Prince

Edward, about 10 miles away, dredged but could not land in the fog, and then headed east toward the Crozets.

Station 147: 30 December 1873

46°16′ S, 48°27′ E; off Crozet Islands, 1,600 fathoms

The ship was at this point crossing the Southwestern Indian Ridge, which continues the Mid-Atlantic Ridge around the tip of Africa into the Indian Ocean. The fog cleared, the sun came out, and two casts of the trawl just short of the Crozets brought up hundreds of animals, among the richest hauls of the voyage, to round out the year.

Grenadiers and cutthroats

There were several fish flopping about in the trawl. Some were grenadiers, *Coryphaenoides*, one of the most abundant of abyssal fishes, found in all the oceans, although here near their southern limit. They are big-eyed, big-headed fish a foot or more in length, with a long tail tapering to a point (their alternative and less grandiose name is "rat-tails"). Others were cutthroat eels, *Histiobranchus*, which are also big-eyed, big-jawed fish with an even more sinuous eel-shaped body. These fish undulate over the sea floor searching for shrimps and sea urchins and smaller fish, and are usually the first animals to arrive at any bait laid out by inquisitive scientists. They also forage far above the sea floor (using their big eyes), eating pelagic fish and squid and, nowadays, fragments of rubber

Figure 34 *Coryphaenoides* from Station 147

Source: A. Gunther (1887), Report on Deep-Sea Fishes. Zoology, Part 52, Plate 34. Bound in Reports, vol. 27.

Figure 35 *Histiobranchus* from Station 147
Source: A. Gunther (1887), Report on Deep-Sea Fishes. Zoology, Part 52, Plate 62. Bound in Reports, vol. 27.

and plastic. Grenadiers supported one of the very few fisheries operating in the deep sea, but they did not support it for long. They could not possibly do so: their lifespan is 20 or 30 years, with a few fish surviving to 70 years of age (roughly the same as human hunter-gatherers). Russian boats (mainly) began to catch them in the North Atlantic in the 1960s, with the usual result that the fishery for these late-maturing, long-lived fish has by now collapsed.

High pressure

Both fish were taken at depths exceeding 2,000 fathoms during the voyage, and spend at least some of their time close to the bottom, where the pressure is 400 atmospheres. (The current depth record for fish is held by a snail-fish, *Pseudoliparis*, photographed at 8,145 meters, about 4,000 fathoms, in the Mariana Trench.) This extreme pressure does not have the same effect on these fish as it would on a human diver, who would be killed instantly, because the pressure is the same inside as outside their bodies. (Recreational scuba divers swim down to about 20 fathoms; the current depth record is about 160 fathoms.) Nevertheless, living tissue is compressible. Earlier in the voyage, Wyville Thomson and Murray had been dumbfounded to see what could happen to a wooden trawl at 2,000 fathoms:

> The beam was broken through the centre, and otherwise most singularly twisted and torn by the great pressure to which it had been subjected; the wood was compressed, so as to reduce the diameter of the beam by half an inch, and the knots projected.

Surface-dwelling fish convulse and die when subjected to this pressure; conversely, deep-sea fish are usually moribund when brought to the surface, even if uninjured by the trawl, and die soon afterward.

There are specific physiological and anatomical adaptations that make it possible to live at great depth, at the expense of being uncomfortable at the surface. The force of muscle contraction, for example, increases with the ambient pressure, but increases much more in deepwater fish like grenadiers and cutthroat eels than it does in surface-dwelling fish; the difference is the degree of specific adaptation to high pressure. There are more subtle modifications too. The biochemical reactions involved in metabolism are catalyzed by enzymes, protein molecules that usually unfold to expose their active site. This causes a slight increase in molecular volume and is therefore inhibited by pressure; deep-sea fish have modified versions of enzymes that are resistant to high pressure. Animals other than fish encounter the same problems at great depth, and adapt accordingly. Gelatinous animals such as medusae, for example, are highly resistant to pressure (because they consist mostly of incompressible water), and deep-sea animals in many groups have independently evolved gelatinous bodies: the record-breaking snailfish has a body that was memorably described as resembling wet tissue paper.

Colossal

The trawl also brought up cephalopods, almost the first from deep water. The webbed octopus *Cirrothauma* was a new species and had probably been caught on the bottom. It is another deep-sea animal with a gelatinous body. Its eye is a simple cup lacking iris and lens, perhaps just enough to detect bioluminescent amphipods and polychaete worms, which it enfolds in the tent of its arms. The little purple squid *Bathyteuthis*, on the other hand, is a visual animal that presumably hunts in midwater and even has unusual accessory eyes, large light-sensitive organs behind its real eyes. It is only about as long as a little finger, but another squid that prowls in these waters is the largest nonvertebrate animal in the ocean. The Colossal Squid *Mesonychoteuthis* is the shorter, fatter relative of the giant squid, *Architeuthis*, which is the most likely candidate for many sightings of sea monsters, at sea or washed up on shore. Adults weigh up to about 500 kilograms, the size of a horse (or if you prefer, a grand piano), and their arms are equipped, not only with suckers, but also with fearsome swiveling hooks. According to some authorities,

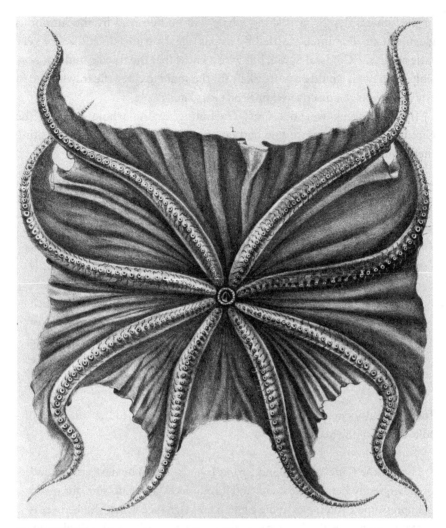

Figure 36 *Cirrothauma* from Station 147
Source: W. E. Hoyle (1886), Report on Cephalopoda. Zoology, Part 44, Plate 10. Bound in Reports, vol. 16.

however, it does not chase down its prey but instead just hangs suspended in midwater to ambush passing fish, which it detects as they brush against its arms. I prefer the fearsome-hunter theory. Its arms are too short for an ambush predator, and its body is muscular rather than gelatinous, as its smaller relatives are. It also has the largest eyes of any animal, a foot across, although it may use them to avoid sperm whales rather than catch its own prey—even in the sunless depths, a rushing whale will be betrayed by the luminous gleams of small animals startled by its passage. In truth, we still know very little about the Colossal Squid, but we do know that the trawl could not possibly catch it—it would be more likely for the squid to catch the trawl—and in any case it was not discovered until 50 years later.

The next week was the most frustrating of the whole voyage. The Circumnavigation Committee had emphasized the importance of visiting these southern islands, especially the Crozets:

> Special attention should be paid to the botany and zoology of the Marion Islands, the Crozets, Kerguelen Land, and any new groups of islands which may possibly be met with in the region to the south-east of the Cape of Good Hope. Probably investigations in these latitudes may be difficult; it must be remembered, however, that the marine fauna of these regions is nearly unknown, that it must bear a most interesting relation to the fauna of high northern latitudes, that the region is inaccessible except under such circumstances as the present, and that every addition to our knowledge of it will be of value.... Two spots [Prince Edward's Island and Crozets] more interesting for the exploration of their vegetation do not exist upon the face of the globe. Every effort should be made to make a complete collection.

The Admiralty's instructions to Nares give more weight to the "may be difficult" aspect of the task:

> Leaving the Cape, Marion and Crozet Islands should be next visited, and subsequently Kerguelen's Land, and it is unnecessary, I am sure, for me to impress upon you the extreme caution and vigilance which will be necessary in navigating this boisterous and little known region with a single ship, even in the middle of the summer season.

As it turned out, no landing could be made with safety on the Crozets. Hog Island, one of the group, was sighted early in the morning of 31 December, and the ship stood in toward land at first light, but the fog was too dense and she was forced back out to sea. The fog lifted a little and Penguin Island was sighted in the late afternoon, but Nares shortened sail in the strong northwest wind and stood off the island for the night in thick fog. On New Year's Day the fog lifted in mid-afternoon, but the wind shifted and freshened and the ship moved out to sea in thick and rainy weather. The next day the wind rose to gale force before moderating in the evening, and Nares then tried for Possession Island. By the evening the ship had arrived off Navire Bay, where there were some deserted sealers' huts, but the swell and the surf prevented any landing—the bay is no more than an indentation in the rugged coast—and Nares, much to the disappointment of the naturalists, exercised the extreme caution demanded by the Admiralty, gave up on the Crozets, and shaped a course instead for Kerguelen.

The next three days were no better and must have been much worse below decks. The ship made good progress running before a northwesterly gale, but was rolling 12 to 25 degrees each side of the vertical, five times every minute, fit to make the figurehead seasick. Sounding and dredging were impossible, work in the laboratory was impossible, and the sick list lengthened. It became very cold, with abrupt squalls of hail or snow, and the ship was repeatedly broached with icy water. But the ship (and crew) held together and reeled off 700 miles in little more than three days to reach the dark, forbidding coast of Desolation Island: Kerguelen.

Station 149: 9–29 January 1874

40°08′ S, 70°12′ E; off Kerguelen, 20–127 fathoms

Challenger steamed past the cliffs of Christmas Harbour, on the northern-most peninsula of Kerguelen, and dropped anchor in 18 fathoms. The site was not convenient as a base, however, and the following day she set out for the eastern peninsula of the island and found a snug anchorage at Betsy Cove, in Accessible Bay, walled in by cliffs. If anything, it was a little too snug—the stern lay only a few feet from the kelp—but it was protected from most sides against the wind. Surveying parties were sent out, and the weather turned so fine that the sailors were allowed to run on shore, their first excursion since the Cape. The very next day, however, the barometer dropped, and a fierce squall swung the ship across the cove and threatened to hammer the stern on the rocks, a serious affair with no shipyard closer than Australia; a skillfully placed anchor saved them from disaster. The next three weeks were spent surveying the eastern and southern coasts of the island, dipping into bays and fjords and sounding and dredging in shallow water, while the naturalists ranged over the gloomy hills. The surveying parties had plenty to do, as Kerguelen had seldom been visited before, except by sealers, and only the Ross expedition of 1840 had mapped it thoroughly. Among other things, they could give whatever names they wished to prominent features of the landscape, and the map soon became littered with the names of *Challenger* sponsors, scientists, and officers: Buchanan Island, Murray Island, and Suhm Island; Balfour Rock; Mount Wyville Thomson, Mount Havergal, Mount Richards, Mount Crosbie, and Mount Tizard; Cape Maclear; the Aldrich Channel; and the Wyville Thomson Peninsula. Most of these names still appear on current French charts (France assumed sovereignty over the island 20 years after the *Challenger* visit and continues to administer it). Willemoes-Suhm was very proud of Suhm Island, and sent a sketch of it to his mother back in Germany.

The peculiar animals

The dredging was extraordinarily productive. After a dozen hauls more than 300 species had been recorded, about 200 of them new. The bays seemed to be floored with sponges, hydroids, tunicates, polychetes, bryozoans, ostracods, and isopods; a whole submarine zoo was emptied onto the dredging platform, time after time. Above all, these seas are the paradise of echinoderms. All the five main groups of echinoderms had already been dredged up in the Atlantic: starfish (asteroids), brittle stars (ophiuroids), sea cucumbers (holothurians), sea urchins (echinoids), and the emblematic crinoids. They can be found more or less everywhere in the sea, but they became more prominent in deep water, and in the Antarctic they often formed the bulk of the catch.

Despite their abundance, echinoderms are very odd animals. If you bring to mind a worm or a snail or a crab you could draw up a catalog of things you expect them to have, a sort of animal parts list. They will have a head, for example, with a brain, or at least a clump of ganglia that serves as one; eyes; a heart and blood vessels; an alimentary canal emptying through an anus; a kidney or its equivalent; reproductive ducts, and perhaps a penis and a specialized female tract; gills for respiration; and so forth. Well, echinoderms have none of these, none at all. Their ancestors had all of them, but they have lost them all, one by one, in the course of the long evolutionary trail they have followed since their origin more than 500 million years ago. Someone (George Orwell, I think) once said that the theme of every Shakespearian tragedy could be summed up in a single word. For *King Lear* the word was "renunciation." Echinoderms are the King Lear of the animal world, the outcome of the Great Echinoderm Renunciation. What they got in exchange was plumbing.

If you look inside the arm of a starfish, for example, you will find gut and gonads, both rather sketchily constructed, and, in sharp contrast, a very sophisticated plumbing system, which is not concerned with drainage but instead supplies hydraulic power for locomotion, attachment, prey capture, and food handling. Like all hydraulic systems, it is filled with water, drawn in from the sea through a plate on the upper surface of the central disk of the animal, something like a perforated manhole cover, and pressurized by cilia. The water is led into a circular canal that runs round the inside of the disk and gives off five radial canals, one for each arm. The radial canal in turn gives off a series of pairs of lateral canals down the length of each arm, each ending in a peculiar structure that looks like the bulb of an old-fashioned bicycle horn. Squeezing the bulb (by muscles in its wall) extends a long flat-topped cylinder through gaps in the skin; relaxing the bulb, while contracting muscles in the wall of the cylinder, makes it retract. This cylinder is the tube foot: applying its flat top to a hard surface produces a pulling force when it retracts, rather like a rubber sucker applied to a sheet of glass. The force exerted by a single tube foot is quite modest, but a medium-sized starfish has upwards of a thousand tube feet, and their combined force could support something the weight of a newborn baby, although they are more likely to be used for pulling apart a bivalve such as a mussel, so that the starfish can evert its stomach through the opening and liquefy its prey with digestive juices. I do not in the least understand why such eccentrically constructed animals should be so successful, especially in the deep Southern Ocean. I am in good company: Libby Henrietta Hyman, the author of a classic series of texts on invertebrates, concluded her volume on echinoderms by writing, "I hereby salute the echinoderms as a noble group expressly designed to puzzle the zoologist."

Cabbage

The hills, in contrast to the bays, were almost barren, except for the dense colonies of nesting seabirds. The flora had previously been described by Joseph Hooker, then the botanist to the Ross expedition and later to become a crucial supporter of Darwin at the time of the publication of his *Origin of*

Species. There were only about 30 species of plant on the whole island. The most remarkable was the Kerguelen cabbage, *Pringlea*, which had been discovered during Cook's visit to the island a century before. It is edible, and the sailors collected it by the boatload, although opinions about its palatability vary: Campbell, who did not like it, says that it gradually disappeared from the wardroom menu, although "the men liked it, and always had lots for dinner." They would certainly be encouraged to have lots, because it is rich in vitamin C and prevents scurvy, the plague of long-distance ocean voyages: Anson, for example, lost two-thirds of his crew to scurvy in the course of his circumnavigation in the 1740s. Cabbage and wild duck, which the officers shot by the score, were no doubt a welcome change for the men from the eternal pease pudding and salt pork. Everyone agreed that it was a most remarkable instance of the inscrutable working of Providence that so useful a plant should be found in so remote a place, exactly where it was needed most.

Transit of Venus

In fact the naturalists were just footling around, while the real work was being done by the survey teams. Shallow dredging and duck shooting were all very well, but would scarcely justify so long a stay. The reason is given in the Admiralty instructions:

> Kerguelen's Land will be a fertile field of exploration in every department of science, and acquires additional interest as one of the stations selected for the observations of the transit of Venus in December 1874. . . . It is desirable that the longitude of the transit station at Kerguelen's Land should be ascertained with the greatest possible accuracy.

A transit of Venus is the rare occasion when Venus, visible as a small black circle, crosses the disk of the Sun. It was prized by astronomers because it offers a way of estimating the distance of the Earth from the Sun. The method depends on the familiar phenomenon of parallax. Traveling in a train, for example, objects in the foreground appear to move relative to the background, so that a farmhouse, say, that is lined up with a church steeple on the horizon at one point will be lined up with a wooded hill a little further on. If you know the distance you have traveled and the angular separation of steeple and hill, you can calculate the distance to the horizon by trigonometry. In the same way, Venus will appear to follow a different path across the sun when viewed from distant localities, and the difference can be used to calculate the fundamental astronomic unit, the distance between earth and sun. The calculation requires knowing the relative distances of earth and Venus from the sun, but this is provided by Kepler's third law, which relates the period of a planet's orbit to its distance from the sun. More practically, it involves calculating a very small angular distance from observing a small wobbling blob move across the roiling surface of a distant star. The Admiralty hydrographer Richards, who was instrumental in planning the voyage, was skeptical about the prospects of success: "We are told to sail to the Antarctic Continent and to visit a variety of small rocky islets interspersed over the Southern Ocean. . . . many of which are actual myths, while on those which do exist it is certain that there is no anchorage for a ship, and that even landing would be generally impossible." He might have added that, if an island did exist, and a landing could be made, the appointed day would probably be cloudy.

Despite Richards's doubts, a party was sent to Kerguelen and arrived there in early October. The men spent the next two months choosing sites, setting up equipment, and praying for fine weather on The Day. Since a clergyman, Reverend S. J. Perry, was in the party, their prayers were efficacious, and 8 December dawned fine and clear. Except that, as Reverend Perry recalled:

A little later we began to feel somewhat nervous, as a dense cloud, not very large in extent, had made its appearance, and was travelling slowly but steadily in the direction of the sun. When the latter became at last completely eclipsed by the cloud, we were delighted to find that it wanted several minutes to the most important instant of internal contact. We hoped that there would be ample time for the small cloud to pass off. We waited, uneasy but not discouraged. The slightest breeze would have saved us, but the wind, usually so extravagant in its generosity to the Kerguelen hills, was now silent and niggardly, and the dreaded cloud seemed to settle down, as if in its destined place, and there remained covering the sun for fully twenty minutes.

Fortunately, the combined efforts at three observing stations did yield some useful results, and the work of the survey team was over, or so they thought. It came as a surprise to them that the Astronomer Royal, Sir George Airy, had (rather airily) given instructions that they were to remain there to make lunar observations in order to fix their absolute longitude with certainty. One hundred double observations of lunar altitude were required, in fact, which given the state of the skies was no welcome prospect. In the event it was not until the end of February 1875 that they could dismantle the huts, pack the equipment, and leave the island, by which time they were down to half-rations and rueing the day they had taken up astronomy. As Reverend Perry remarked, "To be a martyr for science is all very well in contemplation, but all may not find it so agreeable in the practice."

The officers continued surveying (and naming) until the end of the month, when the ship sailed from Christmas Harbour on a cloudy morning, rounding Cape Digby in the east before steering southeast for Heard Island. As they cleared the land, they fixed the position of the southernmost cape and named that too: Cape Challenger.

Station 151: 7 February 1874

52°59' S, 73°33' E; off Heard Island, 75 fathoms

The weather was fine enough on the morning of 2 February for the dredge to be used on a gravel bottom at 150 fathoms, bringing up dozens of animals and a load of small irregular dolerite boulders. In the afternoon, however, a dense fog came down, forcing Nares to reef sails and proceed with great caution: the position of Heard and the McDonald Islands was somewhat uncertain, and there was now the risk of encountering icebergs. Soundings were of no help, as the rocky bottom rose and fell irregularly, and for the next two days the ship groped her way uneasily to the southeast. Without a sun sighting, Nares then decided for safety to sail south to the latitude of Heard, then steer east along the line until they found the island. On 6 February the weather cleared enough to sight the two small McDonald Islands through the mist (all these uninhabited southern islands were named for the captains of the ships that sighted them first) and soon afterward anchored in Whisky Bay, on the north shore of Heard Island.

Nares, Buchanan, and Moseley landed on a beach of smooth black sand at the head of the bay, where they were met by half a dozen sealers with rifles in their hands. These were hard, rough men who lived in primitive huts, little more than a sheet of wood over a pit dug into the ground, and burned penguin skins for warmth. They might pass years on the island, trekking over glaciers to the exposed beach, impracticable for any boat, where the sea elephants breed. They thought that the *Challenger* had lost her way, being unable to imagine any other reason for her appearance in so remote a place; the truth—that the men they met earned a handsome fortune, by sealers' standards, by dredging up worthless starfish—would have been altogether beyond their imagination. The beach was backed by a sandy plain thickly strewn, like an old battleground, with thousands of the skeletons of fur seals and sea elephants. Beyond that was only barren land with no green plants, rather like Kerguelen but without the charming climate and the picturesque landscapes of Desolation Island. There was no time to explore it; Moseley had only enough time to snatch up a few souvenirs from the charnel heaps before the ship signaled for the shore party to return. There was a heavy sea running, and early the next morning, as the snow began to fall heavily, Nares decided to leave the island rather than attempt another landing. As the ship pulled away from the shore, Moseley caught a last glimpse of one of the sealers, already pacing the shore of this desolate place, littered with the bones of his victims, where he lived the most brutish of lives in a filthy hovel, separated for years from women and kin, living on scanty stores and the acrid flesh of penguins, and all for a few barrels of oil and an armful of stinking skins.

Cold water

An hour or so later the ship stopped off a small island a few miles away to the north. The wind had dropped almost completely, and despite the thick fog the scientifics used the opportunity to dredge in 75 fathoms over volcanic sand. The dredge came up filled with animals, the same rich variety they had found off Kerguelen. "In these cold regions," Murray remarks, "the bottom of the sea seemed to be teeming with animal life." It was cold indeed. The seas close to the coast of Antarctica are so cold that they freeze, forming sea ice and icebergs. Since ice is pure water the rest of the sea is made a little saltier. This cold salty water flows out from Antarctica until it meets slightly warmer subantarctic water, and being more dense dips beneath. This is the Antarctic Convergence, which roughly speaking lies a little to the north of the Antarctic Circle: Marion and the Crozets lie outside, Heard lies inside, and Kerguelen is more or less astride. The cold bottom current formed in this way flows eastward along the coast of Antarctica before swinging north into the Pacific, where it is warmed and rises to the surface. It then flows south, between Australia and Indonesia, round the tip of Africa and north into the North Atlantic, where Arctic ice reverses the process and sends a deep cold current southward off the east coasts of North and South America toward the Antarctic, where it completes the loop. This is the "conveyor belt" that helps to shape the overall dynamics of the world ocean. It is a very slow conveyor belt—a packet of water might take a thousand years to travel from one end to the other—but it has a powerful influence on marine life. In particular, it ensures that deep water is oxygenated, except in enclosed basins like the Black Sea, and can support animal life.

The trawl brought up a lot of animals. Less expectedly, it also brought up a boy's boot, which was a difficult object to explain. It might have fallen from an Australian liner, but these ships sailed far to the north, suggesting, perhaps, that a deep current flowed toward the Antarctic in this region. This and other learned speculations were cut short by the coxswain. One of the ship's boys had left at the Cape, leaving behind a boot that the coxswain had tossed overboard that morning, on the reasonable supposition that Willie would never appear to claim it. The animals were also interesting. As before, there were lots of sponges, polychaete worms, peracarid crustaceans (isopods, amphipods, and others), large pycnogonids, bryozoans, and echinoderms among the animals in the dredge. Other groups, however, were poorly represented, or even completely missing, although they had been prominent at previous stations. There were no bivalves, for example. There were not even any decapod crustaceans—crabs, lobsters, and the kind of shrimp you buy at the fishmonger—although these had been very abundant in the Atlantic. On the other hand, there were, besides the others, crinoids and brachiopods, so the community as a whole had an antique flavor, somewhat reminiscent of Paleozoic seas. Was there, after all, some truth in the old idea that ancient lineages, long extinct elsewhere, had found a refuge in the abyss?

Warm water

There has been deep water ever since animals first evolved, of course, but whether animals have always been able to live there is less certain. Poking around at Christmas Harbour, Moseley had found fossil tree trunks on Kerguelen, where there are certainly no forests now. Scott's expedition later found fossil wood on the mainland, but these observations were so puzzling that they were widely ignored until the weight of evidence slowly made it clear that forests had indeed grown in ages long past both in the Arctic and in the Antarctic. In the Mesozoic, at the time of the dinosaurs, the world was warmer and the Antarctic continent, then as now close to the South Pole, supported extensive coniferous and broadleaf forests. Toward the end of this period these changed to temperate forests as the land cooled and glaciers began to form in the mountains. By the Miocene, about 14 million years ago, the forests had been replaced by tundra, and soon afterward even this was obliterated as the ice sheets advanced. A few shreds of fossil bark and leaves and the occasional piece of a trunk, covered by the ice, are all that now remains of the vast forests that once covered the southern lands. What we got in exchange was the conveyor belt that ventilates the deep sea, whose motive power is the cold dense water of the Antarctic coast. In warmer times, this conveyor belt, deprived of its motive power, would stop, and with it would stop the circulation of oxygenated water in the deep sea. Any oxygen remaining would soon be used up by respiration, the deep water would become anoxic, and the animals would all die. The deep sea cannot have been a refuge for Paleozoic animals, then, and the present-day communities can be no more than about 14 million years old.

But perhaps this conclusion is premature. There were at least two long ice ages during the Paleozoic that would have fostered life in the deep sea. There were even some deepwater communities during the long afternoon of the Mesozoic, although we know little about them: the rocks formed by deep-sea sediments do not often outcrop on land, so the fossil record is thin, but it does exist. Perhaps the animals retreated foot by foot to the summits of seamounts close enough to the surface to remain habitable. Or perhaps there were a few places where local circulation refreshed the deep water. Or perhaps we should be thinking in terms of cold-water rather than deepwater communities—the animals caught off Kerguelen and Heard Island were from shallow water, after all. The hypothesis is not quite dead yet.

Salt water

The crew collected samples of seawater from fixed depths at each sounding, and these were carefully analyzed by Buchanan in the chemistry lab, presumably during reasonably calm weather. Ocean currents are driven by differences in temperature and salinity, so the thousands of measurements that he made during the voyage were later used to understand the principles of ocean dynamics. To a layman, perhaps, the sea is more or less the same almost everywhere, wet enough to drown you but too salty to drink. Indeed, its chemical composition does not vary greatly from place to place, at least in the open ocean: mostly sodium chloride, but with substantial amounts of magnesium, sulfate, calcium, and potassium. Oddly enough, chemical analyses of the body fluids of marine animals show that they often have a similar composition. How can this be explained?

Many of the animals collected during the expedition had simple cup-shaped bodies, with a large central cavity closed by a mouth. Anemones, corals, jellyfish, and hydroids all follow this fundamental plan, although they are often covered with elaborately sculptured tissues. The central cavity is necessarily filled with seawater, which can be circulated by the ciliated cells that form its wall. Captured prey are pushed in through the mouth by the tentacles and digested there, with the remains expelled by the same route. The cavity thereby serves for irrigation, respiration, digestion, and waste removal. Other animals, which diverged during the great radiation of animals hundreds of millions of years ago, have more or less elongate, worm-shaped bodies built around a central tubular gut, open at both ends. This makes digestion much more efficient, but it also poses a problem, because fresh, oxygenated water can no longer be circulated within their bodies because they no longer have a central cavity. If they are sufficiently small, or sufficiently flat, diffusion will do the job, but large, bulky animals built on this plan were impracticable until a secondary fluid-filled body cavity had evolved, an interior space between the body wall and the gut wall. But what would the working fluid be? The only feasible candidate was seawater, because for long ages past cells and tissues had become adapted to working in a seawater medium, and natural selection never invents but only modifies. This is why the internal body fluids of marine animals, serving for respiration, waste disposal, nutrient transfer, and storage, are chemically similar to seawater. Some animals, like echinoderms, have a body cavity open to the sea and use seawater itself as a working fluid. If they were placed in fresh water, it would flood their bodies, killing them as their cells swelled up, which is why starfish don't live in ponds.

In most animals the secondary body cavity is divided into compartments. Annelids and arthropods take this to an extreme and have segmented bodies divided up into many compartments separated from one another by sheets of tissue. This arrangement makes it possible to evolve very complex, highly differentiated bodies, because each segment can be independently modified (to bear an antenna, say, or a walking leg) without needing to remodel the entire body. It creates the same problem, though: the bulkheads between each compartment prevent the circulation of fluid through the body. This drawback has been overcome by the evolution of blood vessels and ducts that pass through the barriers that separate the segments and allow nutritive, oxygenated blood to supply tissues along the whole length of the body. If we ask, what is the working fluid? the answer is the same: seawater. It is not exactly the same as seawater; quite apart from proteins and other organic substances, most internal fluids are less concentrated (which is why you can't drink seawater) and have a different chemical balance than seawater, with much less magnesium, for example, but the general resemblance is unmistakable. The first amphibians that crawled onto land took all of this apparatus with them, and their descendants retain it still. Blood serum, lymph, pleural fluid, seminal fluid, amniotic fluid; all of these are based on dilute seawater to which a complex broth of proteins, sugars, antibodies, hormones, and so forth have been added. The depths of the sea may seem to be a strange and hostile place, but that is where our ancestors lived, and we carry the memory of it around with us every day. The sea is in our blood.

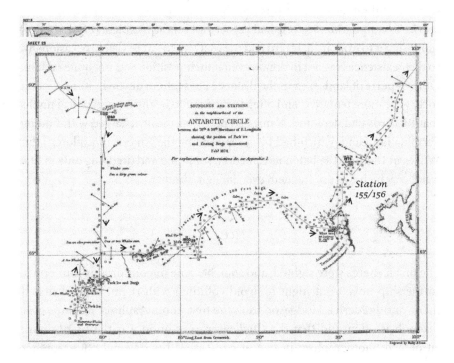

Unnamed station between 155 and 156: 24 February 1874

About 64° S, 95° E; continental slope of Antarctica, depth unknown

Heard Island cared nothing for the *Challenger* and slapped her in the face as she departed. A rough sea smashed two ports and flooded the sickbay, which fortunately was untenanted. The next few days were fine and clear, however, and it was even possible to use the dredge, although it came up twisted, with just a few animals inside. One of them was a new fish, *Bathydraco*. Like all Antarctic fish, *Bathydraco* has somehow to overcome the difficulties of living in very cold water. At the surface, a herring or a haddock would freeze solid, and the icefish that live under the ice can survive only by losing their red blood cells and secreting antifreeze into the plasma, so that their blood is colorless and resists freezing, like radiator fluid. In the deep sea the conditions are a little less extreme: the freezing point of seawater is depressed by the high pressure, and ice crystals do not form, but *Bathydraco* nevertheless needs some antifreeze, although it has red blood.

The main task for this section of the voyage was not to catch fish, however, but to search for land to the south and east of Heard. The geography of the southern seas was still rather vague at this time because the ice was such a formidable barrier to exploration. No one had yet set foot on the mainland of Antarctica, supposing it to exist at all. Heard and the McDonalds were visited occasionally, so their existence was not in doubt, even if their position was not quite certain. Other reports of land, seen in the distance but without anyone setting foot on rock, were more tentative, and often turned out to be the topographical myths that Richards had deplored. Nares had been specifically charged with finding Termination Land, which had been reported by the American explorer John Wilkes in 1840, and sailed in its direction, sounding and dredging only in fine weather, which was not frequent even in high summer.

Ice

The first icebergs were sighted, and soon became uncomfortably numerous, so the ship hove to at night to avoid colliding with them. Wild sketched them, and Frederick Hodgeson took the first photographs of icebergs ever made. It is often said that the *Challenger* expedition was the first to have an official photographer; in fact, it had three. The original photographer was Caleb Newbold, a corporal in the Royal Engineers, which had run a training course in photography for several years, mainly to contribute to surveying work. He ran at Cape Town. Hodgeson was found in a hurry by Wyville Thomson to replace him, but he later ran at Hong Kong. The third photographer was Jesse Lay, who remained for the rest of the voyage. The three of them took hundreds of photographs, but almost all of them show scenes on land, because photography at sea was practicable only in very calm conditions.

The pack ice was sighted on 14 February, a white bar along the southern horizon, and two days later *Challenger* became the first steam-powered ship to cross the Antarctic Circle, at 66°34′ S, sailing through thick, brash ice. Tizard tried to land on an iceberg (to make magnetic readings clear of the iron of the ship), but the large bergs were unscalable cliffs of ice and the smaller ones moved too much to be comfortable. The pack was not visible, but Nares feared being trapped by the ice and tacked north in a sea full of whales, which were safe for the time being, when the whaling ships did not venture further south than Kerguelen. The sea was also full of icebergs, which evidently made tempting targets, and on one occasion the ship closed with one and fired a light gun at it. The temptation was irresistible—opportunities for live firing were rare—but perhaps unwise, as several tons of ice were dislodged by the shot and cascaded into the sea between berg and ship, "as if," Campbell said, "the whole berg were coming down about our ears." "We are too close," Wyville Thomson cried out, "we are too close!" but the ship only rocked about a bit, and the gunner put another shot into the berg, higher up, which simply made a hole in the snow.

The early morning of 24 February was calm enough to put the dredge over, but it had to be hauled up quickly—empty - as the wind began to freshen from the south, and soon reached gale force, with snow beginning to fall heavily. With steam up, Nares moved into the lee of a large berg to cut the wind, a normal maneuver, and sent men aloft to reef the topsails. What happened next is not entirely clear; perhaps the wind dropped and there was too much way on the ship; at all events she struck the berg and scraped along its side, smashing the jib boom. The masts held, but the men on the fore topsail yard, only a few feet from the ice cliff, must have had a bad fright, and they rushed down the rigging to the deck. A sailing ship would have been helpless, but the puny engine, put into reverse, slowly dragged *Challenger* clear of the berg. She was still in serious trouble, with the weather thickening and the wind rising. There were icebergs all around, but visibility was less than 100 yards and speed had to be maintained to give steerage way. It was intensely cold, with the wind hurling the snow into the faces of the men working to clear the wrecked headgear "like dried peas." Sooner or later an iceberg must lie directly in their path and there would be little time to avoid it. This then happened; Nares and Maclear bellowed orders from the bridge, the engine was put to full speed astern, the topsail backed, and a vast dim white mass loomed up—and passed. The ship drifted on in fog and snow until another berg was sighted ahead, and Nares could use the safe space between the two bergs to shift backward and forward between the lee of one and the lee of the other, until the gale subsided during the night, when the officers of the watch could breathe a sigh of relief and the captain could go below after 12 hours or more on the bridge.

Termination Land

The ship sailed through loose stream ice close to the reputed location of Termination Land on the day following her collision with the iceberg, and found nothing there. Two days previously she had sounded in 1,300 fathoms, which seemed to settle the matter. In any case, James Clark Ross, in *Erebus*, had sailed over the same patch of sea in search of Wilkes Land some 30 years before. Yet Wilkes, despite a checkered career—he was court-martialed twice—had presumably seen something resembling land. A distant bank of cloud can look very solid. Or perhaps it might have been a very large, flat-topped berg, or ice shelf, that had grounded on the seabed? No berg grounds in 1,300 fathoms, of course, but a tongue of very shallow water extends from the mainland 100 miles to the south, and there is a reliable Soviet report from the 1960s of a berg some 40 miles long, Pobeda Ice Island, just off its northeastern flank. The Antarctic was a vague region in the nineteenth century: large tabular bergs can form, shift, melt, and disappear, so that what appears to be solid land in one year is replaced by deep water the next. Murray himself thought that the mainland could not be far way, and he was right, but it proved to be elusive, and even today there are very few places where it can be approached directly by ship. Richards was right to be skeptical, but it is difficult not to be sympathetic toward these old mariners, groping their way through the bewildering maze of ice, anxious to be the first to find solid ground on the elusive continent. Anyway, Wilkes had the last laugh. He gave his name to Wilkes Land on the coast of Antarctica opposite the supposed location of Termination Land; extended as a sector to the South Pole, it is about the size of Western Europe.

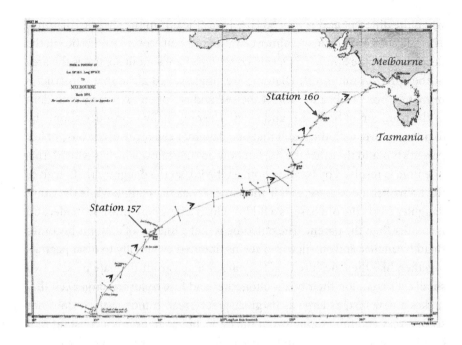

Station 157: 3 March 1874

53°55′ S, 108°35′ E; South Indian Basin, 1,950 fathoms

Challenger now sailed northward from the coast in a gale of wind and heavy snow squalls, but she was sailing away from the ice, and the number of bergs in sight decreased from day to day. By 1 March the number was down to single digits and Nares thought it safe to sail through the night. This was very nearly fatal, as they narrowly missed a large berg in the middle watch, so narrowly that the sails flapped under its lee. A few days later they saw their last berg, a mere melting lump the size and shape of an upturned ship, and were clear of the ice for good. The day before, 3 March, the wind had been light enough to send down the trawl, which came back up full of mud and animals; both were interesting.

Diatom ooze

The ooze of the Atlantic seabed had consisted mostly of calcium carbonate from the shells of forams and pelagic snails. The ooze from recent dredges was quite different and consisted largely of silica from the shells of diatoms. This siliceous ooze forms an almost continuous belt around Antarctica in the deep water of the Southern Ocean, covering an area about six times the area of the Amazon rainforest. Diatoms are single-celled algae that construct a miniature box of silica to protect the cell and are often extremely abundant in the plankton of both seas and lakes. They normally reproduce simply by dividing in two, with the result that one daughter cell receives the base of her parent's box and the other the lid, but how do they know which is which? This dilemma is resolved by treating both as the lid, so the daughter that receives the lid makes a new base, and is the same size as the parent, while the other daughter treats the old base as a lid for which a new base must be made, and is smaller than the parent. The outcome is that a lineage of diatoms becomes steadily smaller and smaller, and seems doomed eventually to disappear altogether, like the Cheshire cat. They are rescued from this fate by sex: two small cells abandon their boxes altogether and fuse to form a larger cell that makes a new box, as large as its great-great-great-grandparent. It does not seem a very carefully thought-out life cycle, but at all events a constant rain of minute silica boxes falls to the ocean floor wherever there are diatoms at the surface. The other plankton organisms that use silica, and contribute to the ooze, are the radiolarians, which make fantastically intricate skeletons to protect an amoeboid creature that captures other plankton (including diatoms) with long thin filaments of cytoplasm. Oddly enough, both diatoms and radiolarians make their skeletons out of the amorphous form of silica called opal, so the bed of the Southern Ocean might be thought of as an enormous flat semiprecious stone. It looks a lot like mud, though.

Spun glass

The third source of siliceous ooze is sponge spicules, which are often made of silica (also opal, as it happens), and make a little heap of silica whenever a sponge dies and decomposes. These heaps add up; there may be a layer a meter or more thick of spicules in the sediment. Glass sponges, the hexactinellids, are particularly common in the Antarctic, although *Challenger* found them in deep water all over the world. They are a little difficult to describe because there is nothing similar they can be compared with. It would be an exaggeration to say (although some have said it) that animals can be divided into two main kinds, the hexactinellids on the one

Figure 37 *Euplectella* from Station 157
Source: F. E. Schulze (1887), Report on Hexactinellida. Zoology, Part 53, Plate 5. Bound in Reports, vol. 21.

hand, and everything else on the other. Still, they are very peculiar animals indeed. Their bodies are supported by a meshwork of fused spicules, so that their cleaned skeleton is a thing of ethereal beauty. The "Venus's flower basket" is a hexactinellid (*Euplectella*, collected by *Challenger* from several stations) that was used as an ornament, mounted under a cloche glass, in Victorian drawing rooms. The body itself is constructed differently from other animals: rather than being made up of separate cells, it is mostly a continuous thin sheet of cytoplasm coating the spicules, with nuclei scattered within like currants in an Eccles cake. It is as though a very complicated scaffolding had been dipped lightly in egg white; this is a good design for filtering bacteria out of seawater, but it is too fragile to withstand waves or strong currents, so they live in the calm waters of the deep sea, where they may grow to the height of a man. Very little is known about them, because they are generally dredged up damaged and dying, and it is only recently that we have been able to see them in place on the seabed through remotely controlled cameras. They are certainly ancient, going back to the dawn of animal life, above half a billion years ago. They may also live for a century or more; some even speculate that large individuals are thousands of years old, the oldest animals on Earth. I doubt it; they seem to have few defenses against the incessant nibbling of starfish and amphipods.

A fierce sponge

Hexactinellids were very abundant on the sea floor off the Antarctic coast, but there were even more species of their lumpier relatives, the demosponges. These are smaller versions of the barrel sponge collected the previous September in the equatorial Atlantic, and make a living in the usual sponge style, by pumping in seawater through pores in the body wall and expelling it through a large opening at the top, the osculum, filtering out bacteria in chambers lined with flagellated cells on the way. After the voyage, the demosponges were sent to the British Museum, where they were described by Stuart Ridley and Arthur Dendy (Dendy was prominent enough to be remembered by a photograph in the National Portrait Gallery). One of the specimens collected at Station 157, a sponge called *Cladorhiza*, puzzled them. It was undoubtedly a sponge, and yet it seemed "to be without oscula or pores, nor have we succeeded in finding flagellated chambers . . . it seems just possible, therefore, that . . . these sponges have some method of obtaining their supplies of nutriment that is quite different from that found in other sponges; this is, however, extremely unlikely." *Cladorhiza* had a long stalk

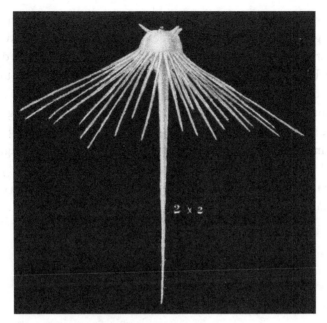

Figure 38 *Cladorhiza* from Station 157
Source: S. O. Ridley (1887), Report on Monaxonida. Zoology, Part 59, Plate 20. Bound in Reports, vol. 20.

bearing a body set with a whorl of long filaments, something like a parasol. According to Ridley and Dendy, the function of the filaments "is doubtless to prevent the sponge from sinking into the soft mud on which it lies." How could they possibly have guessed the extraordinary truth? The filaments are covered with minute hooked spicules (think of Velcro) that trap the setae (minute hairs) of any copepod that brushes past. The animals struggles, but only to be snagged by more spicules. Arthur Ransome somewhere describes a similar tactic used by the Volga fishermen to catch sturgeon. They lay out long, light lines bearing unbaited hooks on the riverbed; a sturgeon pricked by one will roll or rub to remove the irritant, only to be hooked by more, until at last it is so enmeshed that it can be drawn up on shore. In *Cladorhiza*, more filaments begin to grow toward the prey, until it is wrapped up completely and can be digested by wandering cells. *Cladorhiza*, in brief, is a carnivorous sponge, the least likely of animals.

Perhaps we should not be so surprised; after all, carnivorous plants are well known, and perhaps even less likely. The two unlikely predators even seem to evolve along similar lines: besides hooked spicules, other carnivorous sponges use sticky traps, like sundews, or suction traps, working by the collapse of a water-filled chamber, like bladderworts. I've already described how the bivalve *Rhinoclama* has turned from filter-feeding to predation by evolving a suction trap. Why should deep-sea animals be so fierce? Carnivorous plants use the insects they trap as a source of nitrogen that is otherwise scarce in the bogs where most of them live. There is no shortage of nitrogen in the deep sea, but there is, of course, a perennial shortage of food in general. Lying in wait for a passing copepod may require a lot of patience, but there is nothing else to do, and it saves the energy needed to operate a water pump continuously.

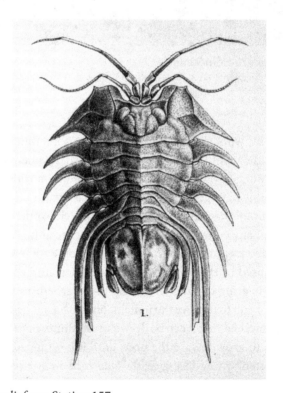

Figure 39 *Serolis* from Station 157

Source: F. E. Beddard (1884), Report on Isopoda. First, Part: Genus *Serolis*. Zoology, Part 33, Plate 4. Bound in Reports, vol. 11.

False witnesses

Trilobites are the emblematic animals of the Paleozoic, appearing near its beginning, in the middle Cambrian, and disappearing at its end, during the mass extinction that closed the Permian. They were mostly fairly large animals, flattened dorsoventrally, with large compound eyes looking upward borne on a head shield, and two rows of legs fringing their thorax. Trilobites would have sculled over the seabed using their thoracic legs, eating any small soft-bodied animal they came across, and peering fearfully upward for anything hoping to eat them. Nothing could have given a stronger boost to the notion of the deep sea as a refuge for ancient animals than a trilobite crawling out of the dredge and into the view of an astonished Willemoes-Suhm. Something tantalizingly like a trilobite was indeed found in several hauls. It was a fairly big animal (well, a couple of inches long) with a flattened oval shape, a head shield with sessile compound eyes, a fringe of walking legs along its thorax, and a movable tail shield—just like a trilobite. Willemoes-Suhm was not deceived, identified it as a large, deep-sea isopod, and, in the best *Challenger* tradition, named it *Serolis bromleyana*, after Lieutenant A. C. B. Bromley. One would like to think that his heart skipped half a beat, though.

Along with trilobites and hexactinellids, the early Cambrian fauna included an unusual group of mollusks called monoplacophorans. They are unimpressive animals the size of a fingernail, looking something like a limpet, and lived uneventful lives until disappearing toward the end of the Devonian period, about 360 million years ago. No trace of them could be found in younger rocks, and they were very reasonably assumed to be extinct, until to every zoologist's astonishment a living monoplacophoran was dredged up from the Puerto Rico Trench by the Danish survey ship *Galathea* in 1952. Even better, it seemed to be segmented, with a series of paired gills, kidneys, and muscles on each side of its body. This was nothing like a limpet, and caused a furor in zoological circles, where words like "segmented" tend to bring out the worst in people. Monoplacophorans are not common, but they have since been found in several other deepwater sites, including the Southern Ocean, and they are just the sort of small sedentary animals that could readily have been swept up by the *Challenger*'s dredge or trawl, and would have been the crowning discovery of the voyage. But they never were, and the naturalists remained unaware of the strange survivor that they never caught.

Station 160: 13 March 1874

42°42′ S, 134°10′ E; South Australian Basin, 2,600 fathoms

Challenger continued north and east toward Australia, crossing the Southeastern Indian Ridge into deeper water. The seabed was changing to a more familiar pattern, passing from the diatom ooze of the Southern Ocean to globigerina ooze and finally, in very deep water, to red clay. The fauna was changing too, as the Antarctic receded and warm-water forms began to appear. The sea began to be luminous again at night, and a *Pyrosoma* was caught by the trawl, presumably close to the surface, on 10 March. This was the spectacularly luminous colonial tunicate that had fascinated the officers last June, off Newfoundland. The trawl also contained an odd amphipod called *Phronima*, which had puzzled naturalists by swimming around in a gelatinous shell, quite unlike any other amphipod. The solution turned out to be theft: the animal creeps into the tunic of a young pyrosome and sticks its legs out at the back so that it can propel it through the water while it slowly consumes the rightful owner. A few days later, the trawl reached red clay and brought up a variety of interesting objects from the deep sea floor, including the black pebbles that Buchanan had previously shown to be made up largely of manganese. Attached to one of them was what appeared to be a small plastic bag about the size of a matchbox, which since it sported two siphons was clearly a tunicate, although of an unusual kind. It turned out to be yet another new species, which was eventually named *Abyssascidia wyvillii* in honor of Wyville Thomson, who by now was accumulating a lengthening list of objects named after him, from tunicates to mountains. Its tunic was thin and almost transparent, with a rather stiff consistency because (as was discovered long afterward) it was made largely of cellulose, which is perfectly scandalous, although to explain why will need a slight digression.

The enigmatic tunic

We are eukaryotes, with bodies made up of cells with nuclei. So is everything else except bacteria and archaea, the prokaryotes. Eukaryotes arose through some kind of fusion—the details are still hotly debated—between different kinds of prokaryote and diverged very early in their history into two main branches, one with a single posterior flagellum (usually) that evolved into modern fungi and animals, and one with two flagella, one at the front and the other at the rear (usually), which evolved into everything else, including seaweeds, land plants, and a host of less familiar organisms. Our lot, the one-flagellum eukaryotes, use chitin as a structural fiber, for example in the cell walls of fungi, the cuticle of nematodes, the exoskeleton of arthropods, and the radula of mollusks. The other lot use cellulose as a structural fiber, especially in the cell walls of plants. Chitin is tough; cellulose is stiff, which is why trees are possible. Bacteria can produce both, but eukaryotes for the most part are restricted to one or the other. Cellulose would be very useful to animals, but they do not have the enzymes necessary to produce it; natural selection is very effective in modifying structures, but cannot by itself create anything wholly novel, so no cellulose-based animals have ever evolved. Except tunicates. To everyone's surprise, tunicates—all tunicates, not just *Abyssascidia*—incorporate cellulose into their outer covering, the tunic, and they have the genes to prove it. But where did they get them from in the first place? This is a serious problem, because the obvious answers—that they alone invented cellulose genes, for example, or that they are not really animals at all—range from the unpalatable to the unacceptable. It was resolved only when new technologies made it possible to sequence the cellulose genes and find out where they came from. They are unlike any other animal gene, but they are very similar to cellulose-synthesizing genes in bacteria. Bacteria exchange genes quite freely (this is how multiple-drug-resistant strains of pathogenic bacteria originate), but eukaryotes do not, although on very rare occasions they may acquire new genes from bacteria. Tunicates can synthesize cellulose because long ago in their evolutionary history they accidentally incorporated a bacterial gene into their genome, and have since been able to make good use of their unexpected gift. It follows, of course, that all tunicates are genetically modified organisms—GMOs—despite bearing no warning label.

The weather cleared, and the decks dried for the first time in six weeks. A month earlier, Swire had written "The men, of course, have to live on salt grub, but we have sheep still left, besides any quantity of preserved meat and vegetables and wine, etc, in unlimited quantity." By now there were no sheep left and even the officers looked forward to the comforts of Australia. The ship had been flung about for weeks in rough seas that often made hot meals infrequent and laboratory work impossible. The icebergs that had been an intriguing novelty when they were first seen in mid-February had become a bogey by early March, and Matkin didn't care if he never saw an iceberg again, except in a picture. Some of the men were so apprehensive that they stayed up all night, bad enough in itself, but worse considered as a comment on the state of discipline on the ship. The *Challenger* had succeeded in surveying, sounding, and sampling in Antarctic waters but had been bruised by the experience and was glad to limp into Melbourne on the last of the coal, mooring in Hobson's Bay on 17 March 1874. Campbell remarked, "There was joy among us on arriving at Melbourne."

12

Seventh Leg

The Coral Sea, June–September 1874

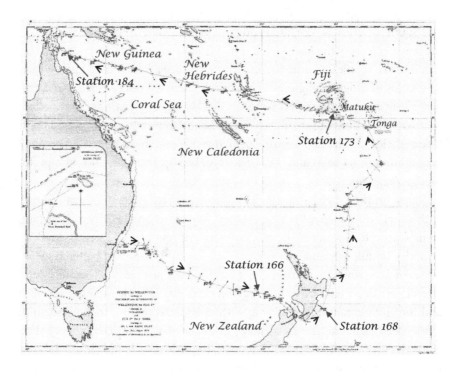

Challenger spent two weeks in Melbourne, where incoming swell makes the anchorage awkward, before shifting to Port Jackson, now known as Sydney, one of the best natural harbors in the world, on 6 April, where she stayed for the next two months. She needed a refit after being bumped around in the Antarctic (although the damage was not as serious as had been feared) and had to take on enough coal and supplies to last till Hong Kong. There was also a lot of work involved in packing the specimens from the Southern Ocean into 65 large boxes and 10 casks, to be dispatched to Edinburgh. Nevertheless, the long stay in Sydney had something of a lackadaisical air. The country was neither as rigorous as the southern islands nor as exotic as Brazil; for an excursion

Full Fathom 5000. Graham Bell, Oxford University Press. © Graham Bell 2022.
DOI: 10.1093/oso/9780197541579.003.0013

you could take a bus or a train, using one of the free passes given to the expedition by the rail company, and stay in a comfortable house with fellow countrymen, observing (or, if you were Moseley or Campbell, shooting) the native fauna. Moseley took the bus to Browern Creek to see a living platypus, but had to be content with buying a couple of dead and somewhat decomposed specimens. Willemoes-Suhm traveled to Mount Macedon and was attacked by land leeches. The most ambitious trip was made by Wyville Thomson, Moseley, Murray, and Aldrich, who spent three weeks in northern Queensland searching for the Australian lungfish, *Ceratodus*. This animal, with its relatives in Africa and South America, is very interesting indeed to zoologists because it is one of the few survivors of the lineage that is the sister group to the tetrapods—all the amphibians, reptiles, mammals, and birds. The area was so little known that new kinds of lungfish might even be found (there are none, in fact), and the party went equipped with nets and even, as a last resort, dynamite. They netted or blasted hundreds of fish, but no lungfish, until, just before they were about to leave, Aldrich caught one on a worm. But there seems to have been no particular hurry to get back, and with *Challenger* laid up, there were no attempts to dredge along the coast.

Nares, on the other hand, was a seriously worried man and went around with a dark countenance because the men were deserting in droves. They had some reason to. Several of the officers were unpopular, and knew it: Matkin remarks acidulously that the officers "are treating the men better now in hopes of deterring others from going but the men just laugh." The men were also dissatisfied that they were not given extra pay for cold-weather service, and the endless rain filled the ship with mosquitos that bit mercilessly and made sleep difficult. The root cause, however, may have been simply that Australia was a land of opportunity, with English speech and (more or less) English customs, where they could easily abscond and, perhaps less easily, make their fortunes. Six men simply used their train pass for an excursion to Ballarat, to see the goldfields, and failed to use the return half. Others were less fortunate, or just less intelligent: two stole one of the ship's boats to get to shore but were inevitably captured and jailed. By the middle of May, 22 men had run and five were in jail. No wonder Nares furrowed his brow.

A precious clam

Willemoes-Suhm found the most interesting animal just by poking around in a sandbank at the entrance to the harbor. It was a rather clumsy-looking bivalve, with a strong ribbed shell about the size of a walnut, called *Trigonia* (now *Neotrigonia*). It would have been immediately recognizable to anyone who has hunted for fossils in Jurassic or Cretaceous ironstones or sandstones, where thousands can often be found. When the dinosaurs disappeared, *Trigonia* disappeared too, although it did not subsequently inspire as many documentaries. It did not completely disappear, either, because a few species managed to survive along the southern coast of Australia, where they are neatly adapted to high-energy environments swept by surf or currents. They have a large strong digging foot, with a sort of heel, so strong that they are capable of making a rather feeble but unexpected jump. To extend the foot, the valves of the shell must gape widely, but have deep ridges on the inside to lock the valves securely together when the shell closes again, a little like a ziplock bag. The outside of the shell bears knobs and ridges that grip the sediment as the animal burrows. Altogether they seem well equipped to live in inshore environments, where they are often thrown around or buried, and why they have been almost universally replaced by feebler thin-shelled bivalves in recent times is not clear. The inside of the shell has a beautiful deep pink luster, and was used to make earrings for the ladies of Sydney; Willemoes-Suhm collected a dozen and sent them back to his mother.

A shark with molars

The naturalists knew that another interesting animal lived in the harbor, although they never captured one (they do not seem to have been trying hard at this point). This is the Port Jackson Shark, *Heterodontus*, which looks more like an enormous gudgeon than a shark. PJS does not behave like a typical shark either, lying on the bottom and chewing clams and starfish. Most sharks can do neither. In the first place, they use ram ventilation, swimming with their mouth open to force water over their gills. This seems like a good two-for-one deal, but has the drawback that they suffocate if they stop swimming. A shark like PJS that spends most of its time lying on the seabed must evolve some means of actively pumping water over its gills in order to breathe. Most sharks cannot chew either, of course, being notoriously slash-and-gulp predators that tear off pieces from their prey with sharp serrated teeth that work like a battery of steak knives. PJS has two kinds of teeth (hence its Latin name), neither like a steak knife: pincers at the front are used to pluck hard-shelled animals from the bottom and flat plates at the rear to break them open and grind them. The bite force of its "molar" teeth is about 300 newtons, which, as it happens, is just about enough to crack a walnut. This is much more powerful than the bite of a dogfish of similar size (only about 20 newtons). We use about 100 newtons in normal chewing, but can generate 1,000 newtons or more with our molars if required; the champion biters are alligators and crocodiles, which should be avoided because they close their jaws with a force of about 15,000 newtons.

Although very different from typical sharks, PJS looks very similar to animals known as fossils from the early Jurassic, about 180 million years ago, so, like *Trigonia*, it is often called a "living fossil." This is not an entirely straightforward term. If it is used to mean only that some animal bears a strong superficial resemblance to a long-extinct ancestor, then it is harmless enough. Most "living fossils" have another feature in common, however: they all have more prosperous relatives. The sister group of PJS (together with half-a-dozen very similar species) includes most other sharks: carpet sharks, nurse sharks, tiger sharks, hammerheads, basking sharks, and, of course, the great white shark. You can see why PJS feels like a poor relative. All the other living fossils that I've mentioned—*Trigonia*, monoplacophorans, platypuses, lungfish, and even stalked crinoids—are likewise groups with many fewer species than their closest relatives. This imbalance might be explained in any number of ways, but the implication of the term "living fossil" is usually that such animals have retained primitive features and so resemble the common ancestor of themselves and their more prolific sisters. This would be a valuable clue to working out how a particular group has evolved, but it is likely to be entirely misleading because, as a moment's reflection will make clear, the time that has elapsed between this common ancestor and either the living fossil or any representative of its more numerous sister group is exactly the same. It is as though your family had two branches, one of which had blossomed into a horde of cousins, aunts, and uncles occupying several postcodes in Toronto while the other had been reduced to a maiden aunt living in Moose Jaw. Any pattern of accidents in birth, marriage, and death might account for the discrepancy, but you will not be tempted to assume that your great-great-great-grandparents looked like the maiden aunt. (Purists will object to this analogy because it assumes, improbably, that your family reproduces asexually, but never mind them.) The conclusion might be justified by photographs (or fossils), but otherwise the inference is insecure.

The officers finally dragged themselves away from Sydney, where they had been very well received. They expressed their gratitude by throwing a grand ball for the notables of the town, a sumptuous and glittering affair, although the ship's band was smaller than usual because several of its members had deserted. Nor were the men themselves forgotten. They were invited to a tea meeting, followed by a temperance lecture in the Masonic Hall.

Station 166: 23 June 1874

38°50′ S, 169°20′ E; Challenger Plateau, 275 fathoms

Challenger was seen off by a large crowd and a musical serenade from the bands of the ships in the harbor when she left Sydney for New Zealand on 8 June, 25 men short of complement despite retrieving several deserters who had been jailed. The main purpose of the passage was again surveying, this time to take soundings along the line of a proposed telegraph cable linking Australia and New Zealand. Dredging would be attempted when possible, but natural history was secondary to the main business to find out whether the seabed was suitable for cable-laying; ideally it would be covered by a thick layer of ooze, and any hard bottom should at any rate be smooth (which would in fact best be shown by seeing whether the dredge snagged). During calm weather in shallow water it was quite straightforward to take a sounding—essentially, lowering a heavy weight on a rope until it reached the bottom—except that a sailing ship would inevitably move with wind or current as the lead descended, until the sounding line was far from vertical and the depth consequently uncertain. I have already mentioned how *Challenger* could take accurate soundings by using her engine to balance wind and current with the screw, holding the ship almost stationary in the water. In preparing to sound, then, steam was first got up, the sails furled, and the ship turned head to wind. The leadsman stood on one of the two sounding platforms that jutted from the side of the ship at the level of the foremast on the upper deck and lowered the weighted sounding line, which led from storage reels on deck through a pulley suspended from elastic "accumulators," directly into the sea. The sounding line itself was a hemp rope about as thick as a little finger, with a breaking strain of about 1,500 pounds or 700 kilograms, marked at 25-fathom intervals with a color-coded twist of worsted. The weights were up to eight 56-pound cast-iron cylinders, which detached automatically from the line when the bottom was reached.

Their combined weight drove a hollow iron cylinder closed with a butterfly valve into the sediment, taking a sample that could be hauled back to the surface and preserved. Just over 500 samples were taken like this during the voyage, and although one jar of mud is much the same as the next to the untrained eye, they revealed, perhaps, just as much about the deep sea as the animals captured by dredge or trawl.

The weather, however, was anything but calm. Even on setting out it was so bad that Nares abandoned the attempt, since sounding was out of the question, and returned to anchor in Watson's Bay. The ship finally struggled clear of land on 12 June and set out east across the Tasman Sea. Within two days they were sounding in the deep water of the Tasman Abyssal Plain. This needed more sophisticated equipment than the simple deep-sea lead, but it was lost anyway when the line broke. The men attributed the failure to working on the Sabbath and (more plausibly) to having no issue of surveying wine recently. The wind increased to a gale that reached force 10 the following day, raising a heavy swell with 20-foot waves breaking over the deck, smashing the ship's cutter, and most of the men's crockery. Once the wind had abated, another sounding was attempted, but the line broke again ("supposed to have been cut," Balfour states darkly). Curiously, this was the last time during the voyage that the sounding rope broke; perhaps the wine issue was reinstated.

Venus' girdle

The wind then died away altogether, leaving a calm that lasted a couple of days and obliged Nares to proceed under steam. The dredge was put over in 2,600 fathoms, but retrieved only a scanty haul from the red clay: a few ophiuroids, a small shrimp, and some odds and ends. The sea was so smooth, indeed, that Murray was able to take the boat out and use a tow-net to sample animals at the surface. One animal he saw, but did not catch, was a very long, golden-green ribbon undulating through the water, almost certainly the Venus' girdle, *Cestum*. This is a very peculiar kind of ctenophore, a group of animals that has been mentioned before but not described. They have a unique propulsion system. Broadly speaking, animals less than about one millimeter in length use cilia to move through the water, because bands of cilia provide a continuously operating low-power motor that works well provided that you are very small. Above about one centimeter, cilia are too feeble, and active locomotion requires muscles to work structures like limbs or fins, or simply to wriggle like an eel. Ctenophores are able to ignore this rule because they have evolved the unique trick of gluing together cilia into rows of stiff blades, the so-called comb rows, which act like banks of miniature oars to propel an animal the size of a gooseberry through the water quite fast. They are all predators able to catch copepods and other plankton using long tentacles armed with sticky harpoons. *Cestum* takes this further by growing into a ribbon-like animal up to a meter in length and about five centimeters wide that can move both by the comb rows and by a sinuous undulation of the whole body. It is sometimes held responsible for reports of the sea serpent, but I am skeptical: a meter is impressive for a ctenophore but modest for a monster.

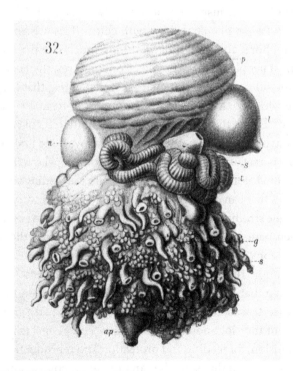

Figure 40 *Stephalia* from Station 166
Source: E. Haeckel (1888), Report on Siphonophora. Zoology, Part 77, Plate 6. Bound in Reports, vol. 28.

Double-bagged

The calm did not last; before long a breeze sprang up and freshened to a gale, although sounding was still possible. The water shelved, first to 1,975 then to 1,100 fathoms, and finally, on 23 June, to only 275 fathoms, despite the land still lying more than 100 miles away. The ship had discovered a broad submarine plateau, in places only about 100 fathoms deep, extending north and west of New Zealand; it is now known as the Challenger Plateau. The trawl was put over twice, and, for the only time on the voyage, provided enough large shrimp not only for the specimen jars but also for the wardroom table. The decapods had returned. The naturalists sifting through the ooze also found a gelatinous blob about an inch across, which the great German zoologist Ernst Haeckel later called "one of the most splendid discoveries of the *Challenger*." He exaggerated; but it really was a most remarkable animal, and if anything has turned out to be even more remarkable than it first seemed.

When the animal was inspected more closely in Haeckel's lab in Jena, most of it consisted of a gas-filled bag lined with chitin. This was enough to place it as a siphonophore, although it did not look much like the only siphonophore that most people might recognize from pictures, the Portuguese Man O'War. Its fanciful name refers to the large inflated bag that keeps it afloat, blown by the wind like a sailing ship (although I don't know why it should be Portuguese). Below the float dangles a rather tattered collection of bulbs and bladders strung together, a little like a broken washing line. Immediately below the float is a set of flask-shaped structures that expel water rhythmically to push the animal along by reaction. Below them are feeding structures with armed tentacles that can deliver a paralyzing jolt to the unwary swimmer, or even to someone strolling along the beach who idly turns over a stranded animal with their bare foot. Some kind of jellyfish, then, with the wicked little poisoned harpoons of its kind, although looked at more closely, its anatomy seems oddly lumpy. This is because each of its unit structures—including the float—is a polyp that has been so highly modified as to be unrecognizable except, of course, to a zoologist like Ernst Haeckel, who had a grand (but mistaken) theory of the siphonophore body plan. Polyps specialized for flotation, for propulsion, for feeding and protection and reproduction, all knitted together into an animal that is neither an individual nor a colony, but somewhere in the debatable land between the two.

Siphonophores are odd enough, but *Stephalia*, the animal in the dredge, is odd even by siphonophore standards. Nothing like it had been seen before, and it was not easy to make sense of it. For a start, it has not only a gas bag but a supplementary gas bag too (the gas, by the way, is carbon monoxide). Why does it need such an outsize float? Nor does it have the usual dangling streamers, which are instead coiled into a spiral beneath the float like hair worn in a bun; the only dangling structures are long tentacles. The obvious guess is that it lives at the surface (the trawl must have caught it on its way down or up), buoyed up by its double bag with its body wrapped beneath its float and its tentacles awaiting customers, which goes to show that an entirely plausible hypothesis may be entirely wrong. We now know that *Stephalia* is after all a truly benthic animal that really was captured by the trawl on the bottom. Its long tentacles anchor it to the seabed while its enormous float keeps it suspended well above, with its body neatly tucked beneath to keep it out of the mud. *Stephalia* is not a sailing ship at all: it is a barrage balloon.

Land was sighted two days later in a strong southeasterly gale, and the ship ran into Port Hardy for shelter. The gale was as fierce as ever when she left on 27 June to fight through the Cook Strait, and Nares was, perhaps, being a little reckless in pushing on so fast, under steam and fore and aft sails, but he was no doubt anxious to reach Wellington without any further delay. At noon they were only a few miles out, and, as was customary, everyone went below to eat (the officers to lunch and the men to dinner), leaving only the watch lieutenant and the four men at the wheel, with the leadsman in the chains. In these shallow coastal waters a leadsman, using a handline, is necessary to give the deck party a running commentary on the depth and timely warning of any sudden shoaling. This skilled and responsible task had been entrusted to a steady young seaman, Edward Winton, who stood, in sea boots and oilskins, on the port sounding platform, heaving the sounding lead. Soon after noon, a passing marine noticed him climbing up from the platform into the chains to free the line, which had become wound round the anchor, no doubt blown there by the wind. Almost immediately afterward the ship was jarred by a lump of sea that broke over the bows and spilled the men's dinner into their laps, but the incident was passed off lightly enough, perhaps because land and some degree of liberty were so close. Ten minutes later, though, the same marine, appalled, noticed that the sounding line was still around the anchor while the platform was empty. Within a couple of minutes the engines had been stopped and all hands called from dinner to turn the ship back on her course. A dark object was soon spotted on the surface; but it turned out to be a clump of seaweed. No trace of Edward Winton was ever seen again. In such a sea, weighed down by heavy clothing, he could not have survived for long, even if he avoided being struck by the blades of the screw. Nares abandoned the search after an hour or so and resumed course for Wellington, anchoring at Port Nicholson later that afternoon. The men held a whip-round and raised 20 pounds for the woman back in England who was now Winton's widow.

Wellington was wet, windy, and cold. Nobody seems to have enjoyed it much, and there were not even any desertions. Willemoes-Suhm is as cheerful as ever in his letters home, but confined himself mostly to visiting the museums; Moseley was ill; Campbell writes blankly that "there was nothing to be seen, and less to be done." They stayed there a week, though, because they were obliged to: they arrived to find the Governor in residence, and protocol demanded that they abandon their plan of shifting to Auckland after a couple of days. They left Port Nicholson without regret on 6 July but again had to do a double-take, anchor in Worser Bay for the night, and wait until the following day for the weather to improve. When they finally did set out for the broad Pacific Ocean it was, I think, with a greater sense of pleasant anticipation than ever before: the Atlantic was home water, more or less. The Antarctic, although impressive, had been actively unpleasant, and Australia and New Zealand, while undoubtedly in the Pacific, were scarcely even foreign, at least to British sailors. Sailing east from New Zealand, they would soon cross the 180-degree line of longitude, halfway around the world, and so were, in this respect, heading homeward, while the coral seas, the exotic islands, and a host of imagined pleasures lay before them.

Station 168: 8 July 1874

40°28′ S, 177°43′ E; Hikurangi Terrace, 1,100 fathoms

A deep groove in the ocean floor runs north-northeast from New Zealand for a thousand miles as straight as a knife-cut. The water is more than 30,000 feet deep in places, 5,000 fathoms, deeper than any sounding yet taken by the *Challenger* and almost the deepest water in the world. This groove is the Kermadec Trench and its northerly continuation, the Tonga Trench, which are formed where the Pacific Plate slides beneath the Australian Plate. The movement of the plates heaps up the sea floor to the west into a ridge that runs parallel to the trenches and often only a few hundred fathoms beneath the surface. The plates are moving very rapidly in geological terms, as much

Figure 41 *Bathysaurus* from Station 168
Source: A. Gunther (1887), Report on Deep-Sea Fishes. Zoology, Part 52, Plate 46. Bound in Reports, vol. 27.

as several inches per year, and submarine volcanoes bubble up where they meet. In some places these volcanoes form islands where they break the surface: the Kermadec Islands, Tonga, and Fiji. The *Challenger* was heading for these islands and so necessarily traveling along the line of the trenches, but these were unsounded seas and nobody on board knew anything of this.

The *Challenger* set off on 8 July up the west coast of North Island under steam in (ironically enough) a flat calm. She was passing over the Hikurangi Plateau, a slab of crust off the west coast of New Zealand that is gradually being subducted beneath the island. It slopes downward to the southwest, and the trawl was put over about halfway between the coast and the abyssal plain. The day was fine enough for the naturalists to putter around in the ship's boat while the trawl was working; when it came up and its contents were dumped onto the dredging platform they found it full of animals, besides a lot of mud.

Catch of the day

The catch of the day, it seemed, was a fish whose name says it all, *Bathysaurus ferox*, the fierce lizard of the deep sea. It was nearly two feet long and looked rather like a small pike, with a boot-shaped snout and a more or less cylindrical body ending in a wedge-shaped tail. Like the pike, it is an ambush predator, designed to lurk and dash: it lies on the seabed, resting on its pectoral fins, with the front of its body slightly raised, gazing into the blackness to detect the least hint of luminescence—its eyes are quite large, and normal in form, although it also has a backup system in the form of a very elongate pectoral fin ray that it may wave around to sense movement nearby. Like the pike, too, it has a mouthful of sharp teeth, although it improves on the pike by providing most of them with a ball-and-socket joint, so that when they bite they collapse downward and inward, preventing their prey from wriggling free. They are probably the top predator of the deep sea, eating other fish like *Synaphobranchus* and halosaurs. Close encounters of any kind must be rare in the deep sea, though, and it uses an enormous liver, nearly a quarter of its body weight, to store the energy from its infrequent meals. At least it cannot itself be eaten by anything else; the deepwater skate *Bathyraja* is found in this area and might try, but it is not much larger and in any case seems to prefer polychaetes. The only other danger is a trawl pulled by a ship, but it is probably agile enough to avoid the clumsy gear of the *Challenger*, and this unlucky fish was the first of only three bathysaurs caught during the voyage.

Figure 42 *Periphylla* from Station 168
Source: Narrative, vol. 1, Part 2 Figure 201.

Pink paint

There were many other animals, though: gigantic ostracods (gigantic at least by ostracod standards—most of these bivalve crustaceans, which swarm in ponds and streams as well as in the sea, are only as big as a hempseed, but these were almost half an inch across); a deepwater nemertean worm, the distant cousin of the terrestrial nemertean that surprised Willemoes-Suhm on Bermuda; an unusual deep-sea jellyfish, *Periphylla*; polychaetes, pycnogonids, mollusks, and the usual range of echinoderms, including a bizarre holothurian, *Enypniastes*. Its most obviously odd feature is its color—it is bright pink rather than the dingy brown of its relatives. It does not loll around on the bottom, either, like other holothurians, but instead uses a webbed cluster of tentacles to swim up off the seabed so that it can move from one feeding site to the next. This is not entirely safe, because it might be intercepted by a predator, but it has a striking solution to that problem: strong contact with another organism triggers intense luminosity, while the delicate skin peels off and drifts away as a glowing pink cloud. Any predator that is fooled into pursuing this phantasm finds that the skin is also very sticky and paints the attacker with bright pink patches, advertising its presence to every other sighted animal in the vicinity.

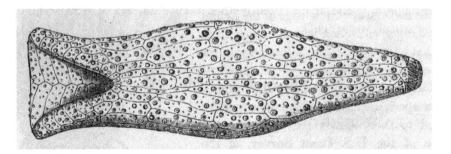

Figure 43 *Pourtalesia* from Station 168
Source: A. Agassiz (188), Report on Echinoidea. Zoology, Part 9, Plate 28a. Bound in Reports, vol. 3. Also in Narrative, vol. 1, Part 1 Figure 82.

Shape-changers

Even ordinary holothurians are a little bizarre, though. The most familiar echinoderms, like starfish and echinoids, the sea urchins, have an obvious fivefold radial symmetry and forage mouth-down for large prey. To evolve a typical holothurian you must imagine a sea urchin rolled onto its side, losing its solid test and then becoming elongated into a sausage shape with its mouth forward, eating mud. It is still pentaradial, with three rows of tube feet below and two above, while its mouth is surrounded by a cluster of five feeding tentacles, but its elongate body gives it the appearance of bilateral symmetry. The common ancestor of echinoderms and chordates was bilaterally symmetrical, evolved radial symmetry in the echinoderm lineage, then re-evolved at least a semblance of bilateral symmetry in the holothurian lineage; even the most fundamental aspects of body plans can be modified in the course of evolution. Another animal caught at this station was an elongate, bottle-shaped echinoderm with only a very thin transparent test and a terminal mouth into which it shovels sediment using spatulate feeding structures. It certainly looks like a holothurian, but it is undoubtedly an urchin, *Pourtalesia*, belonging to an unusual group of deep-sea echinoids. They have evolved a close resemblance to holothurians, including the resumption of bilateral symmetry, because they have taken up a similar way of life based on grazing the upper layer of sediment. Their true nature—how they are related by descent to other groups—is revealed by details of their anatomy (their feeding structures, for example, are modified spines rather than modified tube feet), but their general appearance shows how body plans were not created for the convenience of systematists, in order to provide an easy way of distinguishing one phylum from another, but instead evolved as adaptations to particular ways of life.

Station 173: 24 July 1874

19°09′ S, 179°41′ W; off Matuku Island, 310 fathoms

Challenger pressed on northeast, passing the Kermadec Islands, heading for Tonga. Nares was anxious to make up for the delays and spent little time dredging or even sounding. It was an unfortunate decision. On 15 July the bottom was hard ground at a depth of 600 fathoms, on the ridge; two days

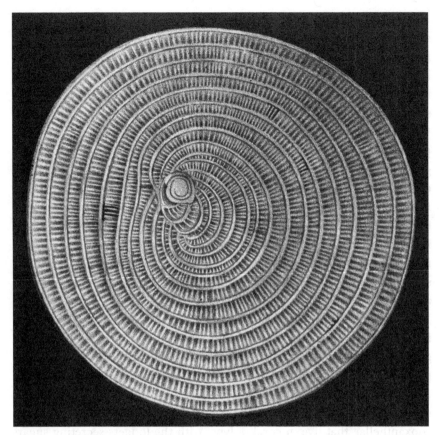

Figure 44 *Orbitolites* from Tonga
Source: W. B. Carpenter (1883), Report on Orbitolites. Zoology, Part 21, Plate 1. Bound in Reports, vol. 7.

later it was red clay at 2,900 fathoms, on the other side of the trench; the next sounding was in very shallow water off Tonga. The expedition had failed to find some of the deepest water in the world, despite sailing over it. It was not discovered until 20 years later, when HMAS *Penguin* recorded 5,000 fathoms in both the Tonga and Kermadec Trenches.

Coins

The ship anchored off Tonga on 19 July and stayed there for three days without any serious zoology being done. Moseley did make one interesting observation, though, as he strolled around the island: the floor of the shallow tidal pools was thickly strewn with white disks about the size of a fingernail, as though someone had tossed in handfuls of small coins. He knew what they were because he was familiar with nummulites, whose fossil remains, cemented together, form a coarse limestone. Each disk is a single-celled foraminiferan, belonging to the same group of organisms as those responsible for the globigerina ooze on the floor of the deep sea. These particular forams were *Orbitolites*, which, unlike *Globigerina*, actually does grow on the bottom, albeit only in very shallow water; it happens to be especially abundant on Tonga, although it can be found elsewhere in the Pacific. It is, of course, a single enormous cell, something like 10,000 times larger than a run-of-the-mill protozoan, which can function normally because it is divided up into many small compartments, each of which acts as a miniature bioreactor under the control of a single central nucleus. *Orbitolites* shows how an organism can become large without being multicellular like an animal.

Moseley collected a sample of these strange creatures and stored them in a jar with alcohol, but found to his surprise that the preserving fluid gradually turned pale green. He guessed that the green color was chlorophyll, dissolved out by the alcohol from algae living as parasites within the forams. He was partly right: there are indeed green lodgers within the body of the forams, but they pay rent. The arrangement is similar to the symbiotic zooxanthellae that live in coral polyps, with the lodgers secreting sugars produced by photosynthesis while the host, a predator, contributes the nitrogen from its prey. Mutualistic relationships like these are quite common, but the agreement between the partners is not legally binding and can become quite shaky. In some cases the symbiont can be passed on from one host generation to the next, from parent to offspring, and is then likely to be domesticated quite quickly because the interests of the two partners become strongly entrained when a relationship is indissoluble. The mitochondria in our own cells, or the chloroplasts of plants, descend from endosymbiotic bacteria whose relationship with their host cells is so intimate that it has become obligatory: neither can survive without the other. Foram hosts do not have so close a bond with their zooxanthellae: they cannot, because the endosymbionts are larger than their gametes, and therefore must be freshly restocked from the sea by every new generation of forams. I imagine that their mutual support is tempered by mutual suspicion. Corals can expel their zooxanthellae if circumstances dictate, like a surly landlord turning out an unprofitable tenant, and although forams are not known to do this, I bet they do.

The crew got on well with the locals; the men were friendly and eager to trade, and the women, despite the missionaries, were scantily clothed. This was certainly much better than mixing with the squalid sealers of Kerguelen. Men came out to the ship in canoes to sell fresh fruit, vegetables, eggs, and pigs, besides a range of miscellaneous items they thought that sailors might just need, and they seem to have judged correctly because Matkin bought a large bunch of bananas, some shells, a fishhook made from a shark's tooth, a paper collar, and an old tie. Nares was careful, though; the men were allowed a run on shore but had to be back by 9:00 p.m. in order "to protect the morals of the natives," as Murray puts it. To everyone's regret, the ship left Tonga in the morning of 22 July, even though "a longer stay . . . would have been appreciated by Members of the Expedition," according to Murray.

Pearly king

The ship reached Matuku, the first of the Fiji group, on 24 July and landed a small party to explore the island while the others (short straw, I suppose) dredged and trawled offshore in about 300 fathoms. The nets brought up a lot of fish and crabs, but these were outshone by a live pearly nautilus, a spectacular prize. *Nautilus* is another one-of-a-kind, like PJS or *Trigonia*, a cephalopod distantly related to squid and cuttlefish but put in a group of its own. Probably everyone has seen a picture of *Nautilus* in its coiled chambered shell, the only cephalopod that has kept the old mollusk habit of living in a shell. The shell is a common enough find for beachcombers, but the living animal is rarely seen. It is another living fossil, the sole survivor of a group that was much more abundant and diverse in the early Paleozoic, hundreds of millions of years ago, and tantalizingly (but perhaps misleadingly) reminiscent of the ammonites, almost as iconic as trilobites for being extinct animals that zoologists would love to find still alive somewhere. It is an active visual predator, with a battery of tentacles (a modified version of the foot that snails use to creep on) able to grasp a crab and a mouth cavity containing a beak and a modified radula, which snails use to rasp your lettuce, to tear it apart. Its eyes are unusual, operating on the principle of a pinhole camera, but presumably work well enough for catching crabs and perhaps keeping out of trouble. It can move quite fast with the same sort of jet-propulsion system as squid, so the specimen captured by the trawl must have been looking the other way. It is not a deepwater animal, because the chambers of its shell contain gas to regulate its buoyancy and would implode at depths of more than 400 or 500 fathoms. It was put into a tub of water for everyone to watch it bobbing around before being consigned to one of Murray's pickle jars.

The moral of the mudskipper

Moseley had gone ashore, of course, and was watching mudskippers, *Periophthalmus*, hopping about among the mangrove roots. These peculiar fish are a kind of goby that can survive out of water by absorbing oxygen through its skin and the lining of its mouth, while using its stiff pectoral fin as a lever to move around on the surface of the mud, where it catches insects who thought they were safe from fish when the tide was out. It is not very well adapted to life on land and never ventures out of the mangroves, because it has not evolved the key features of truly terrestrial vertebrates: an air-breathing lung, the three-section tetrapod limb, and the shelled amniote egg. The reason is history. In very ancient seas, some 400 million years ago, two lineages of fish evolved, one whose fins were webs supported by long bony rays, and the other whose fins grew from the edge of a long fleshy lobe with an internal bony skeleton. There is no doubt which design turned out to be the more efficient and successful: thousands of species of ray-finned fish collectively occupy every kind of aquatic habitat from the village pond to the ocean abyss, whereas the only lobe-finned fishes are a few species of lungfish, plus the coelacanth. Ray-finned fishes like tuna can cruise in the open ocean faster than the *Challenger*, and seahorses can maneuver delicately in a clump of seaweed close to shore, whereas lungfish can only stumble around in the mud of tropical swamps. On the other hand, the ancient lobefins had the beginnings of a lung, and the axial bony skeleton of their lobed fin could be successfully modified through a series of intermediate forms (known from fossils) into the tetrapod limb. The upshot has been that the descendants of the ancient lobefins include salamanders, crocodiles, sparrows, sheep, and us, whereas the summit of terrestrial life among ray-fins is the mudskipper. Efficiency is the enemy of innovation.

Jaws

Moseley spent no more than two or three hours on Matuku before the ship moved on to Kandavu, where they dredged in the mouth of Galoa Harbour. After a brief stay for surveying they sailed to Levuka Harbour, on a small island off Viti Levu, the largest island of the Fiji group. Moseley went off on another jaunt, learning about kava and cannibals, but also hearing of a freshwater shark that lived in a lake shut off from the sea by a cataract. This would be quite anomalous, as sharks do not live in lakes (although there are sharks in Lake Nicaragua), perhaps because they are too big for freshwater. There are some fairly small sharks, of course, but most are large pelagic predators, and a few of them would quickly clear out all the large fish from a lake, leaving them nothing to eat. Then why must sharks be large? Why are there not shoals of cartilaginous tiddlers in our ponds and streams? After all, there were plenty of small cartilaginous fish long ago, and some even lived in freshwater. One plausible answer is that, despite the novel and the movie, sharks simply don't have the jaws. The jaw apparatus of bony fish is an intricate assemblage of bone and muscle that can be endlessly permuted to build a stickleback or a swordfish or anything in between. Shark jaws are good at what they do, but they cannot easily be modified to do anything else. Sharks have never evolved the delicate extensible jaws that bony fish use to follow so many different ways of life, nor are they ever likely to do so, if only because any initial attempt to do so would be merely a clumsy imitation of some structure that had already been perfected by some group of bony fish. I have not been able to locate the lake that Moseley was told about; it is quite likely that juvenile bull sharks, *Carcharodon leucas* (the same species that lives in Lake Nicaragua), forage quite far up the rivers of Fiji, but that is as far as they are likely to get.

Challenger spent several days in Viti Levu before returning to Kandavu, where she remained in harbor for the next week. I have no idea why Nares was messing about like this. Having dashed up from New Zealand to Tonga, only two stations were dredged or even sounded in the next three weeks, and scientific observations were restricted to amateurish observations of mudskippers and freshwater sharks. The only contribution to oceanography was the attempt to locate Metcore Reef, between Kandavu and Levuka, which turned out to be mythical. Fiji is not even mentioned in the Admiralty instructions (which read "From New Zealand your course will be through the Coral Sea toward Torres Straits") and although the ship took on coal in Fiji, there should have been plenty remaining from Wellington. The crew did not even seem to enjoy themselves as much as in Tonga. Nares had been given discretion, though, to explore the small islands to the south and west of New Zealand and, being well ahead of schedule, may simply have elected to visit Fiji instead. It had not been a very fruitful decision.

Station 184: 29 August 1874

12°08′ S, 145°10′ E; Coral Basin, 1,400 fathoms

Challenger left Fiji on 11 August and headed west through the Coral Sea toward Cape York, the northernmost point of Australia, at the entrance to the Torres Strait. The weather was at first wet and changeable, and two days out the fore topgallant mast (the upper section of the foremast) broke in a squall, which would be unexpected in a fully manned ship with alert officers and an expert crew. (Balfour made the excuse that sail was not shortened because there had been several heavy showers during the morning without any wind.) Squalls and showers continued for another couple of days as the ship sailed westward through a sea swarming with *Halobates*, the sea skater. Landfall was made after a week at Tongariki in the New Hebrides (now Vanuatu), but the ship scarcely paused. An armed boat returned a party of laborers from Queensland to Epi, but there was no secure anchorage, and mutual suspicion seems to have prevented any further exploration.

The sea floor over which the ship was passing is unusually complex. The route first passes through the North Fiji Basin, which overlies a southern lobe of the Pacific Plate whose border is the great slash of the South New Hebrides Trench. It then leads through the chain of the New Hebrides, south of Vanuatu and to the north of New Caledonia, as the ship sailed westward into the abyssal plain of the Coral Sea, before reaching the continental slope and the great fringing reefs of the northeastern Australian coast. As if to make up for Fiji, a line of 10 soundings were taken along the way, roughly one every other day. The Admiralty instructions were quite explicit about the importance of sounding as often as practicable: "If any one of the various objects of the expedition is more important than another, it may be said to be the accurate determination of the depth of the ocean." Murray summarized the results of each series of soundings as a section of the ocean showing how depth varied along the route, although 10 soundings along a route more than 2,000 miles long can give only a very crude image of sea-floor topography. They mapped the deep water of the Coral Basin, but missed most of the other features, including the trench. Despite its crudeness, though, the section is cumulative: subsequent voyages can add further soundings to increase the resolution of the section, until it becomes as detailed as desired. Other ships had taken deep-sea soundings before, but none were as accurate or systematic as those taken by the *Challenger*, which continued to provide the basis for ocean surveys until the coming of sonar a century later.

The sounding line was not a simple depth indicator. It was provided with a variety of measuring and sampling devices, which always included thermometers and water bottles. These spoke to the second priority of the voyage, which was to investigate the physical conditions of the ocean. Every sounding provided water samples from specified depths for Buchanan to analyze in the chemistry laboratory, where he labored on more repetitive and painstaking tasks than Moseley (for example) ever set his hand to. The sounding lead, too, came up with a sample of the sediment that was carefully preserved for chemical analysis and the identification of its diatoms and

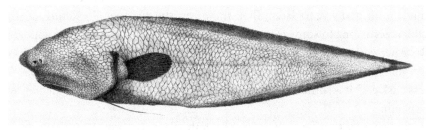

Figure 45 *Typhlonus* from Station 184
Source: A. Gunther (1887), Report on Deep-Sea Fishes. Zoology, Part 52, Plate 25. Bound in Reports, vol. 27.

forams. The measurements of temperature, specific gravity, and dissolved gases of the water samples, and the chemical and biological characterization of the sediments, provided an unprecedented volume of information about the oceans.

The third, and by implication the least important, priority of the voyage, as defined by the Admiralty, was the identity of the animals living in the deep sea. The Admiralty was very interested in depths, currents, temperatures, and even sediments, but it was not really interested in starfish at all, and the dredge was there only because the Royal Society had insisted. However, even the obscure sponges and echinoderms decanted onto the dredging platform held more romance than a table of temperatures, and were certainly the first priority for the naturalists. Unfortunately, the three deepwater dredges in the Coral Sea gave rather disappointing results.

Beauty and the beast

The last of the three, toward the end of August, produced an exquisitely ornamented cone-shaped gastropod aptly called *Basilissa superba*, belonging to a family found only in the deep sea. Since it was less than an inch tall, however, it requires a microscope, and a conchologist, to appreciate it to the full. With it in the dredge was an equally miniature but oddly shaped ophiuroid, *Ophiambix*, that had never been seen before and looks more like a starfish than an ophiuroid, with broad arms and a poorly defined disc. It can fold its arms vertically and wriggle into the holes bored by mollusks such as shipworms in dead wood, where it seems to feed on the wood itself, or perhaps the fungi and protozoans eating the wood, a quite unique diet for echinoderms. A large tree branch also came up in the trawl from this station, and *Ophiambix* presumably fell out of it, although this connection was not made until many years later. Then there was the cusk eel *Typhlonus nasus*, which according to some is the ugliest animal in the world. Its head appears to be perfectly featureless, with no visible eyes, gills, nose, or mouth, so it has the general appearance of a blob with fins and earns its sobriquet of "the face-less cusk eel." It is some relief to find that it has a mouth after all, hidden away beneath.

And that, with a few other familiar animals, was about all; except, per-haps, for a note in the Station-book of a "Patella-like shell" from 1,350 fathoms at Station 175, just after leaving Fiji. *Patella* is the limpet familiar to anyone who has walked along a rocky shore, where it clings onto the rocks by a combination of suction and glue. *Challenger* had found it twice before, but only from shallow water, as expected, since clinging to hard surfaces is only necessary in high-energy, nearshore environments. It is much more surprising to find it on globigerina ooze in deep water. It is not quite clear what the Station-book describes; if it was a shell with the strong ribs of the common limpet, then it was merely a limpet in the wrong place, brought there by some accident. If its shell were smooth, well, monoplacophorans, unknown at the time, look something like a limpet, with a rather flat conical shell broadly open at the base, and perhaps Wyville Thomson missed a major discovery by a whisker. The *Galathea* animal, after all, had been labeled as a limpet and spent years in a vial before someone turned it over and noticed the serial gills. But the *Challenger* shell was lost, or just not kept, and we shall never know.

By 27 August the ship was leaving the Coral Basin and approaching the edge of the Great Barrier Reef at Rapid Horn. She steamed to Raine Island, which sits on the rim of a submarine cliff that extends from the northern tip of Australia to New Guinea and separates the deep water of the basin from the very shallow water of the Arafura Sea. The island itself is a mere dab of sand carpeted with nesting seabirds and, more unusually, a landrail, but a boat landed there to inspect a tall sandstone tower, capped by a rainwater cistern, that had been built there to succor the survivors of any ship wrecked on the reefs. The woodwork of the tower had collapsed and the iron cistern rusted into holes, but Nares did not recommend that it should be repaired, on the grounds that any boat making it this far could just as easily reach the mainland, an opinion that, I imagine, shipwrecked sailors landing there, having seen the tower, might or might not share. Moseley was more considerate and optimistically sowed a small patch of soil with garden vegetables— tomato, pepper, pumpkin, and gooseberry. I doubt that it flourished. Once the landing party had returned, the coral sand offshore was dredged, but for once a dredge in shallow tropical waters was unproductive, yielding only a few bivalves, gastropods, and ostracods of little interest. The crew, meanwhile, were occupied in massacring the sharks that swarmed in these waters; having caught one on the shark hook, they would cut off its tail, slit its belly, and drop it back into the sea, where it would soon be eaten by its companions.

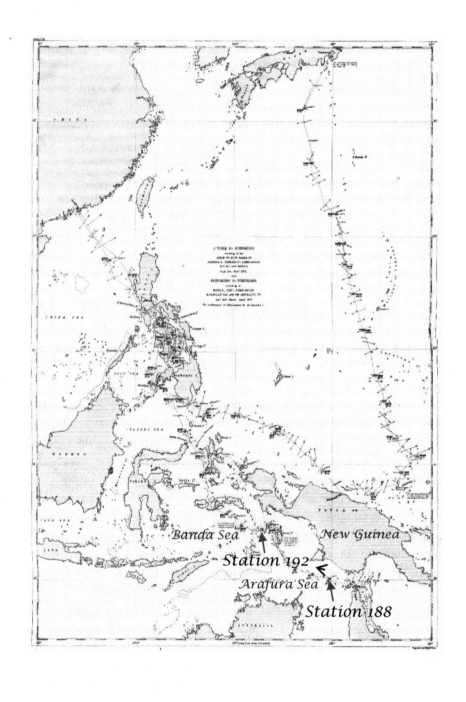

Banda Sea　　　　　　　　*New Guinea*

Station 192

Arafura Sea

Station 188

Station 188: 10 September 1874

09°59' S, 139°42' E; Arafura Sea, 28 fathoms

The mainland was only a day's sail distant, but the way there led through a maze of shoals and reefs, with a strong trade wind blowing over a choppy, uncertain sea. Nares tried to temporize by anchoring on one of the shoals for the night, but they were too small, and when he at length succeeded, the ship slipped off and had to be held on a very long cable to prevent the anchor from dragging on the hard ground. The next day the anchor was weighed as soon as it was full light and Nares himself acted as lookout to guide the ship through the reefs. This was a most unusual procedure, just short of tying himself to the mast, but the trade wind was pushing the ship along at more than 10 knots and the day was misty. Nares threaded his ship through the reefs and along the low barren shore of Queensland before anchoring, not without difficulty, at the remote outpost of Somerset; two or three small vessels were already in the anchorage, so he was forced to lie out in the stream, again on a long cable, with the officers suppressing their irritation.

The lancelet

Challenger stayed at Cape York for the next week, for no good reason that I can find. The officers and naturalists were attracted neither by the country nor by its inhabitants, white or black, nor was there anything very important to keep them there. Campbell disliked the overpowering heat, the impenetrable undergrowth, and the murderous ants. Moseley went out shooting rifle birds and megapodes and brush turkeys; Willemoes-Suhm attacked

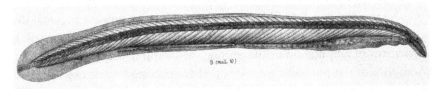

Figure 46 *Branchiostoma* from Station 188
Source: A. Gunther (1889), Report on Pelagic Fishes. Zoology, Part 78, Plate 6. Bound in Reports, vol. 31.

the termite mounds, taller than himself, with a pickaxe to find the queen; Matkin went on a hike, became separated from his friends, and got bushed. Wyville Thomson dredged offshore in very shallow water from the steam pinnace (rather than using the ship itself) and pulled out over 500 species from among the coral. One of them, listed among 30 or so fish, was a very curious animal, the darling of zoologists, the lancelet, *Branchiostoma*. It was long known as *Amphioxus*, until, to the fury of zoologists everywhere, taxonomists announced that *Branchiostoma* had priority, although you may still call it amphioxus if you stick to lowercase roman font. (I remember Bob Carroll instructing our introductory class in its main features by singing "It's a Long Way from Amphioxus" to the tune of "It's a Long Way from Tipperary"; "Branchiostoma" doesn't scan.) It is a mere translucent slip of an animal the size and shape of a large willow leaf. It is undoubtedly a chordate, because it has a notochord, a cartilaginous rod that is the forerunner of the vertebral column, running the length of its body and flexed by blocks of muscle. It also has gill slits, which it uses in feeding by sucking in water through the mouth and expelling it through the slits in its pharynx in order to strain out the edible particles. It spends most of its life buried in sand with only its head showing, sucking in water, but it can swim as fast as a small fish, at least for short periods of time, although since it lacks lateral stabilizing fins, paired sense organs, and a brain, it is not very well equipped for pelagic life. It fits perfectly our preconceptions of what an early chordate would have looked like, a little over 500 million years ago, and there are fossils that seem to fit the bill. In recent years, molecular biologists have tried to spoil the party by suggesting that tunicates are closer than amphioxus to vertebrate ancestry, but sentiment can be stronger than gene sequence, and I suspect many hope that *Branchiostoma* is eventually restored to its place as the sister group of vertebrates.

Challenger left Cape York on 8 September, paused to survey the islets, and landed briefly on Wednesday Island, where they found the giant clam, *Tridacna*, on the reef flats; this is the clam that traps unwary fictional divers who, having accidentally stepped into one, then face the unappealing alternatives of amputation or drowning. The next day they put into Booby Island to visit the post office, a log hut where passing vessels used to leave word of their whereabouts and mail to be forwarded; but there was no mail for them and the ship sailed away to the west. Instead of groping in the dark abyss, *Challenger* was now skating cautiously across the very shallow sunlit waters of the Arafura Sea, sounding every two hours. This was dry land when our ancestors dribbled out of Africa and probed eastward along the shores of the Indian Ocean toward Australia. It was flooded when the glaciers melted 10,000 years ago, and it will be dry land again when they return in a few hundred or a few thousand years from now. Meanwhile, it was a tricky place for a big sailing ship.

Willemoes-Suhm was disappointed with the results of two or three dredge hauls, but they still recorded nearly 300 species; with the dredges off Cape York these were two of the three heaviest hauls of the voyage (the third was off Kerguelen, also in shallow water), in stark contrast to the thin pickings of the deep dredges in the Coral Basin. They illustrated the commonplace observation that the greatest number of species is found in shallow water and diversity falls as depth increases. It might be commonplace, but it has nevertheless been challenged, on the grounds that modern gear such as epibenthic sleds has shown that there are far more species in the deep sea than could ever be captured by dredge or trawl. So there are: but they are mostly small animals. Dredge and trawl captured sponges, soft corals, echinoderms, mollusks, crabs, and fish, all large animals living on or just above the surface of the sediment. There are also many small animals a few millimeters in size, such as cumaceans (comma shrimps) and ostracods, which are able to burrow into the sediment, thrusting aside the detritus. They might be captured by the dredge but would likely be washed out by sieving or simply missed by the naturalists picking over the mud. Others are very small, only a fraction of a millimeter long, in the thin film of water between the sediment particles, grazing on microbes that are smaller still. These minute organisms constitute the interstitial fauna, living in the interstices of the sediment: gastrotrichs, rotifers, kinorhynchs, and other creatures familiar only to the more fastidious kind of zoologist, together with the smallest annelids and flatworms. These would almost certainly be washed out or left unseen; even in the sediment cores their soft bodies would be liable to dissolve in the preservative fluid, leaving no visible trace. These categories are not merely convenient divisions of a continuous range of sizes: they also reflect real physical and biological boundaries. Large animals like echinoderms process sediment in bulk, eating it or pushing through it; they perceive sediment as a viscous fluid. Smaller animals like ostracods handle it as discrete particles that they can individually search or move aside. Very small animals like gastrotrichs do not perceive the sediment as such at all, but must treat each particle separately, since it is bigger than they are. Differences in scale that appear to be merely quantitative can lead to qualitative shifts in the physical conditions, the dynamic regime, that animals experience, and thereby to consistent differences in anatomy and ecology. The sediments are thicker, older, and more stable in the deep sea than in most nearshore environments, and harbor a rich community of small or very small animals that are specialized for living in them. This adds to but does not replace the generalization that large animals become fewer and less diverse in deep water.

A deadly snail

One of the mollusks in the dredge was *Trigonia*, the last specimen found by the expedition; another was *Conus trigonus*, a cone shell, which sounds a little similar but is a very different kind of animal. All snails have a radula, usually a horny pad used to rasp biofilms or plant surfaces into small fragments that can be swallowed. In cone shells the teeth of the radula have evolved into long, sharp, hollow spines filled with venom secreted by a modified salivary gland. These teeth are detachable and can be moved, one at a time, to the tip of a long, extensible proboscis. When it detects a fish, snail, or polychete nearby—each species has its own preference—the proboscis is extended and waved gently around until the animal is contacted. The sharp, barbed radula tooth is then thrust from the tip of the proboscis into the prey, injecting it with venom, while the proboscis grips the tooth firmly to prevent the prey from swimming away. The venom is a potent cocktail of neurotoxins that blocks nervous transmission and sends the prey into convulsions, ending in death. The snail expands its mouth to enormous size to engorge the dead or quiescent prey, and then shuffles back into the sand to digest its meal, slot a new tooth into place, and await its next victim. This is a much deadlier device than the feeble traps of carnivorous sponges or clams. Cone shells are beautifully ornamented and are sought after by professional shell collectors but may be merely picked up idly by tourists, when they will use their sting for defense. If you are stung by a species that habitually preys on worms or snails, you may feel no more than the pain and local swelling dealt out by a wasp, but if you handle a fish-eating species—which include one of the most attractive and desirable species, *Conus geographus*—then the outcome is likely to be muscle paralysis, respiratory failure, and death. Better leave that shell on the beach.

Station 192: 26 September 1874

05°49′ S, 132°15′ E; off Tayandu Islands, 140 fathoms

Challenger crept on, sounding every two hours and dredging from time to time, the sea so shallow that the dredge could be deployed simply by shortening sail, without needing to use steam. The Aru Islands were sighted on 14 September and Nares picked his way very cautiously toward them, noting islands that were wrongly located on the chart and almost discovering an uncharted shoal by running aground on it. The ship finally anchored in Dobo Harbour late in the morning of 16 September and spent the next week there. This was another rather lengthy, unexplained stay. Moseley shot a few birds of paradise, but otherwise little seemed to be done; perhaps Nares was simply filling in time so as to delay reaching the Philippines until the hurricane season was over. The ship left for the Kei Islands on 23 September, immediately finding deeper water. After a brief stay she went on westward to survey a group marked on the Admiralty chart as the Tionfolokker Islands (now called the Tayandu Islands), because little was known about them and even their position was uncertain. It was not even known whether they were inhabited, and the officers might have been tempted to attach their names to the most prominent points, bays, and hills, as they did at Kerguelen, but refrained because "they may already have received names from the Malays, and it would have been a pity to introduce others." The naturalists trawled off the southwest point of Taam and pulled up a rich but rather puzzling mixture of species: many hexactinellid sponges, soft corals, pennatulids, and stalked crinoids, usually indicative of abyssal depths, but also the crabs, cephalopods, and fishes that would be expected in shallow water.

Decorator crabs

There was a splendid haul of crabs, including four new spider crabs—crabs that have long thin legs and a rather globular carapace, more like a harvestman than a spider, really. Their carapace is usually set with sharp spines, large and small, presumably to rebuff predators, although they tend to collect all manner of detritus as the animal picks its way over the sea floor. Some spider crabs make the best of it and pick off any edible bits

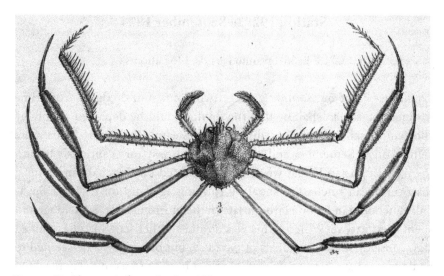

Figure 47 *Platymaia* from Station 192
Source: Narrative, vol. 1, Part 2 Figure 197.

that cling to them, but others have evolved the curious trick of deliberately fastening objects to their carapace, not as a food store but as a disguise. They use their pincers to rip off a piece of seaweed or sponge or bryozoan and can then bend their long spindly legs around to rub it against their carapace, like a man with very long arms scratching his back. These crabs have rows of sharp recurved spines like fishhooks on their carapace, which snag the chosen item and hold it securely. A well-decorated crab looks much like a rock on the seabed—would even smell and feel like a rock—and might well be overlooked by a predator. The decorations are often toxic as well, distasteful sponges (used by the *Hyastenus* captured here) or stinging hydroids (on *Achaeus*, captured previously off the south coast of Australia). Other kinds of crab take decoration even more seriously by cutting out a piece of sponge that fits precisely on their carapace, like a hat on a head, leaving eyes and mouthparts free, or permanently gripping sea anemones locked in their pincers to thrust into the face of a rival or enemy. All the intermediate stages can be found. Spider crabs that live on soft sediment in deep water have little to decorate themselves with (and little reason to do so, in the dark) and have a bare carapace, save for a few encrusting organisms like barnacles, without specialized hooks. Species that live on rocky ground in shallow water, like the lyre crab, *Hyas*, captured the previous year off Halifax in 50 fathoms, festoon themselves with bryozoans and alcyonarians and resemble small boulders mysteriously endowed with movement. Everything depends on circumstance.

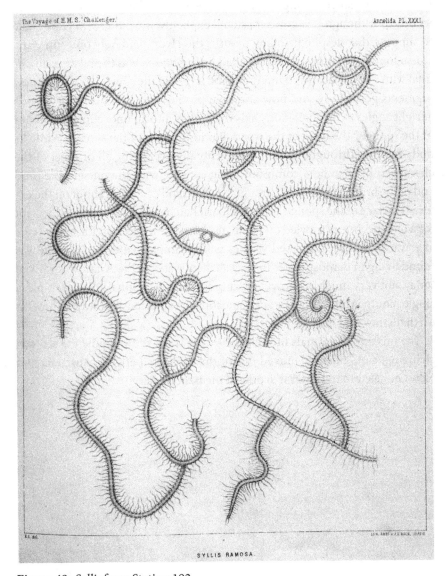

Figure 48 *Syllis* from Station 192

Source: W. C. M'Intosh (1888), Report on Annelida Polychaeta. Zoology, Part 52, Plate 31. Bound in Reports, vol. 24.

The inverse hydra

The most remarkable animal from the trawl, however, was a polychaete that was found inside one of the hexactinellid sponges. In classical mythology, the Lernean Hydra was a many-headed monster that lived in a cave and terrorized the surrounding countryside; Hercules killed it by lopping off all its heads. Although it is nowhere stated explicitly, it would seem reasonable to suppose that despite its many mouths the Hydra had only one anus. The worm found at Station 192, later called *Syllis ramosa*, was the reverse: an animal with a single head and many anuses. It fills up the sponge by adding new segments posteriorly, and now and then producing a new posterior branch at right angles to the main body axis that grows into an unoccupied chamber of the sponge. The head of the worm lies in the base of the sponge, while its body ramifies throughout the host, eventually occupying all or most of the chambers; in chambers communicating directly with the outside the hind end of the branch protrudes through the pore and can move about on the external surface of the sponge. Nothing like it had been seen before, and, except for a few specimens trawled off Japan, nothing like it was seen for many years afterward, until an unrelated species with a similar pattern of growth was found living in demosponges in very shallow water in the Arafura Sea. Even today, not very much is known about these unusual animals. For example, a single mouth is unlikely to be adequate for such an extensive body, so how on earth do they find enough to eat?

The mixture of animals brought up by the trawl showed how the sea was changing, as *Challenger* moved out of the shallows between Australia and New Guinea into the abyssal depths of the Banda Sea.

13

Eighth Leg

The Sea of Islands, September 1874–January 1875

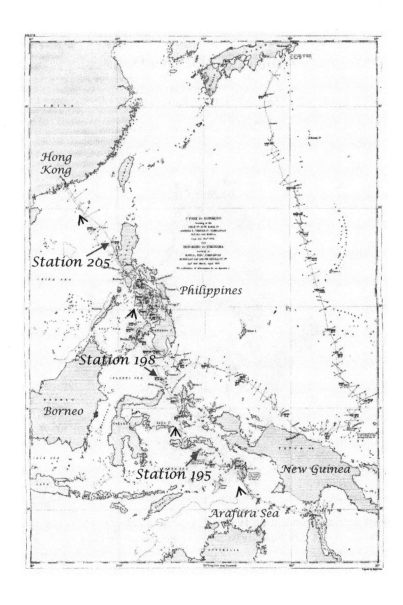

Full Fathom 5000. Graham Bell, Oxford University Press. © Graham Bell 2022.
DOI: 10.1093/oso/9780197541579.003.0014

Two days later *Challenger* sounded in 2,800 fathoms, having sailed over the rim of the Arafura Sea into the huge bowl-shaped depression of the Weber Basin, south of Ceram. Since leaving Cape York she had sailed off the south coast of New Guinea, and would now turn northwest to thread her way past the big islands of the western Pacific on the way to Hong Kong, before looping back along nearly the same track to reach the north coast of New Guinea at the end of February 1875. Much would happen along the way.

Station 195: 3 October 1874

04°21′ S, 129°07′ E; Banda Sea, 1,425 fathoms

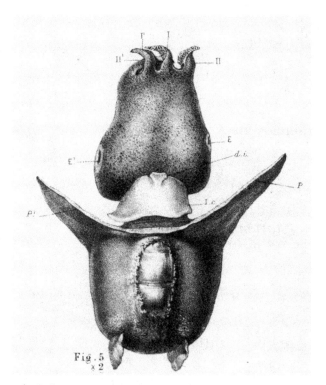

Figure 49 *Spirula* from Station 195

Source: T. H. Huxley and, p. Pelseneer (1895), Report on Spirula. Zoology, Part 83, Plate 1. Bound in Reports, vol. Summary.

Challenger was sailing west, passing over the deepest water of the Weber Basin (which goes down well beyond 3,500 fathoms) before entering the Banda Sea. The Banda Islands were sighted at daybreak on 29 September, but there was time before entering harbor to use the trawl, which to the delight of the naturalists brought up from 360 fathoms a nearly complete specimen of *Spirula*, one of the great zoological puzzles of the time. The shell of the animal was well known, being cast up on tropical beaches by the hundred, but the whole animal is exceedingly rare, so rare, in fact, that the first one found, at Port Nicholson in 1845, was considered too valuable to dissect. It is not particularly impressive, just a small squid-like cephalopod that fits comfortably in the palm of a hand, but its shell is extraordinary: a flat open spiral divided into compartments, rather like a nautilus shell but more loosely coiled and borne inside the body, at the hind end of the mantle, instead of containing the body. Its structure resembles the extinct ammonites and belemnites, so like the nautilus it was regarded as a living fossil and interpreted as the sole survivor of a lineage that branched off very early from the line leading to modern squid. It is poorly known even today; it seems to live in deeper water than most cephalopods and may drift around using the green photophore at the hind end of its body to attract small crustaceans, since it has no beak to deal with larger prey. The specimen in the trawl seemed to be lightly digested and might have been vomited up by one of the fish caught at the same time. The naturalists were justifiably excited; this was the only specimen captured during the entire voyage.

Banda was famous for two sources of misfortune, earthquakes and nutmeg plantations, the one imposed by Nature and the other by the Dutch. A party was taken to see one of the plantations by the local colonial governor, who mentioned in passing that a small steamer had recently broken down in sight of shore and had not been seen since, which was exactly the kind of remark to catch the attention of naval officers. The steam pinnace was promptly sent out and returned a few hours later with the ship's captain and half a dozen of the crew, whom they had found in a small boat, alive but too weak to make for land. They had left the rest of the crew in the ship, the water having run out. Nares had steam raised and took *Challenger* out to search for the ship, but with no success; perhaps it had after all managed to limp into a port on Ceram. There are several points in this narrative that do not seem to ring quite true, but perhaps the captain had abandoned his ship to find help and perhaps the ship had made harbor without sending back word. The pinnace made use of the spare time to

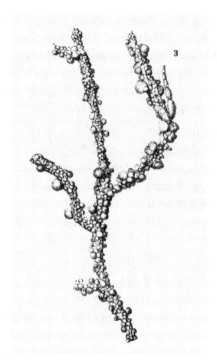

Figure 50 *Rhizammina* from Station 195
Source: H. B. Brady (1884), Report on Foraminifera. Zoology, Part 22, Plate 28. Bound in
Reports, vol. 7.

dredge the harbor, bringing up loads of corals and crinoids but little else; very
few crustaceans, for example. They left Banda the following evening.

The garden of forking tubes

Early the next morning the ship stopped where a depth of 4,000 fathoms had
been reported previously and veered the trawl on 4,400 fathoms of rope, but
on sounding found no more than about 1,400 fathoms, "sufficient proof,"
Campbell sneers, "of the incorrectness of old deep-sea soundings." The trawl
brought up mud with bits of pumice, wood, and coconut shell, and was covered, the Station-book records laconically, "with a tangle of rhizopod tubes."
Similar structures had been found before (on the *Porcupine* voyage, for example), and Murray even states that they were usually seen when the trawl

skimmed the surface layer of sediment. The organism consisted of a net-work of fine brown threads and looked superficially like a filamentous brown seaweed, but cursory examination under a microscope ruled this out, and Henry Brady, who described the material later, christened it *Rhizammina* and identified it as a foram living in agglutinated masses on the sea floor. It was, perhaps, the least photogenic organism found by the expedition, but, after a century of neglect, it might turn out to be among the most important. The microscope shows little more than a light brown branching tube of uni-form diameter (about a millimeter) without segments or compartments or ornamentation. The tube contains dark rounded masses called stercomata because they are believed to be fecal pellets; the cytoplasm is scanty and diffi-cult to stain; the nuclei are elusive. The whole system of endlessly branching tubes forms a tangled clump that often traps sediment and forms a mudball, but the tubes are fragile and without careful handling disintegrate into a soup of short fragments.

   This obscure and uninterpretable organism was then largely forgotten for 100 years, until a group of Russian biologists began to look at it more closely. The thing is that systematic surveys of the deep seabed by box cores often found it, not here and there, but in every sample, so that the top layer of the sediment is a thin living skin consisting chiefly of strands and clumps of *Rhizammina* or something similar. Despite the efforts of some very skillful biologists, however, we still know very little about these cryptic forms of life, and are not even certain what kind of organism they are. The modern brute-force way of finding out is to sequence the genome, but first you have to find the genome, and this has not been straightforward. Simply extracting DNA from the mudball and sequencing it shows any number of genomes from diatoms to dinoflagellates, including annelids, mushrooms, and something close to an onion. These are presumably sequences from organisms that grow on or in the mudballs, or just from spores or seeds that have fallen in; which of them, if any, is the real genome of the creature is impossible to say (although the onion would be a long shot). It is a little like trying to discover the wheat genome by sequencing a plate of spaghetti bolognese. Giving mudballs a formal name is much easier, so they are now Komokiacea, from the Russian for "branching lump," which certainly captures their leading features, although if you want to be familiar you can call them komokis. They remind me of the fungal filaments in terrestrial soils: strange, unseen, omnipresent.

Amboina (Ambon) was sighted the following day, and *Challenger* was moored in the bay that evening, a cable from the pier because the shore is so steep-to. They stayed a week there, partly because coaling was so slow (the siesta time extended from 11:00 a.m. to 4:00 p.m.), and were hospitably received despite the bloody history of the town. Otherwise, not much was accomplished. Moseley went diving on the reef and got sore eyes from staring at fire coral. Campbell and Moseley set out to shoot deer, but were exhausted by hot sun and shoulder-high grass and had no success until, at the end of the day, the beaters drove a doe close enough to the guns to be shot. Or so Moseley says. Campbell's account is different: Moseley had loaded with bird-shot, so that when the deer appeared, he could only pepper them. Campbell ran to cut them off, "the beaters howled horribly, the deer vanished, and intense vexation and irritation against M. reigned supreme." Returning to the ship later in the day, Campbell and Moseley tried to cross the bay in an outrigger canoe, despite warnings from the locals that the sea was too rough and the canoe would be swamped. They set off anyway and had not got far before a wave broke over the canoe and sank it, with only the outriggers keeping them afloat. They were now in the ludicrous position of standing up to paddle an invisible canoe, which required a great deal of effort for very little motive effect. They were now being carried steadily out to sea and had to be ignominiously rescued by the people who had warned them in the first place. One can almost hear the exasperated sighs. Even the shells for which the island was famous were a disappointment; there were plenty of merchants eager to sell them, but the prices were too high for the naturalists, and none were added to the collections.

Station 198: 20 October 1874

02°55' N, 124°53' E; Celebes Sea, 2,150 fathoms

Challenger left Amboina on 10 October and threaded her way northward through the islands. If she had continued west she would have entered the shallow sea between Borneo and Java, little deeper than the Arafura Sea she had just left. Turning to the north, she entered instead the much deeper water of the Molucca Sea, between the Molucca (Maluku) Islands and Celebes (Sulawesi). The ship was now sailing over a tortured region of the crust where two neighboring plates have ridden over the Moluccan miniplate, with the usual consequences of volcanoes, earthquakes, and deep ocean trenches. It is aptly named the Molucca Sea Collision Zone, like an exceptionally accident-prone traffic intersection. She sounded half a degree on either side of the Equator on 13 and 14 October in thousand-fathom water, and anchored a little later in the harbor at Ternate.

The sage of Ternate

The small island of Ternate has a large place in the history of zoology be-cause Alfred Russel Wallace, on his great collecting expedition to the Malay Peninsula, had made his headquarters here between 1858 and 1861—just a decade before serving on the Circumnavigation Committee that dispatched *Challenger* on her very different collecting expedition. He had, unavoidably, chronic bouts of malaria, and it was while (in popular mythology at least) in a mild delirium on the neighboring island of Gilolo (Halmahera) that he con-ceived the idea of natural selection as the means by which new species evolve. He wrote it up a few days later on Ternate with the haste of a young assis-tant professor compiling a tenure dossier and sent the manuscript to Darwin, who realized at once that he had been scooped. (He hadn't been, not quite; Wallace had not thought his idea out thoroughly at this point, but it was close enough to spook Darwin into publication.) The outcome was the joint paper that announced the theory of natural selection to a largely indifferent public in July 1858. The letter itself, surely one of the most portentous in the history of science, has since been lost.

Natural selection was little more than a sideline to Wallace's day job, collecting animals, which led him directly to ideas about the geograph-ical distribution of animals. In particular, he proposed that the Asian and Australian faunas did not merge gradually into one another, but instead switched rather abruptly across a line—"Wallace's Line"—that ran south of the Philippines, continued between Borneo and Celebes, and finally cut the Indonesian island chain through the narrow channel between Bali and Lombok. Plants did not follow the rule as closely, although some (such as *Eucalyptus*) were restricted to Australia until the invention of garden centers. What marine animals did was not clear, but the expedition was intended to find out: the zoological objectives include "the Hydrographic examination of 'Wallace's line' in the Malay Archipelago, and of the littoral faunas on the opposite sides of that line, is of great importance, considering the significance of that line as a boundary between two Distributional provinces." Wyville Thomson, very sensibly, took no notice at all: *Challenger* sailed directly to the Philippines up the west coast of Celebes without investigating the island chain at all. There was no need to; there was no reason to suppose that the littoral fauna (which was not in any case the main point of the expedition) would split across the line, and the naturalists already knew that at greater depths the fauna was divided vertically rather than horizontally.

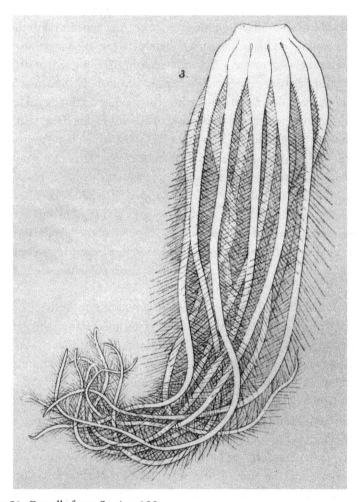

Figure 51 *Freyella* from Station 198

Source: W. Percy Sladen (1889), Report on the Asteroidea. Zoology, Part 51, Plate 115. Bound in Reports, vol. 30.

The cage of thorns

Challenger did not linger in Ternate, leaving on 17 October to cross the Molucca Sea. She rounded the northeast point of Celebes and moved into the Celebes Basin, a large abyssal plain more than 2,000 fathoms deep. On 20 October the sea was calm and most of the day was spent trawling. The net contained some familiar animals, including the ugly fish and the enigmatic xenophyophorans. It was becoming clear that many deep-sea animals were widely distributed and took no notice of Wallace's Line. One was *Parapontophilus*, a large deepwater shrimp that the naturalists had found thousands of miles away in the Atlantic. Another was a distinctive deepwater starfish, *Freyella*, which was found at a dozen or more stations throughout the world during the expedition. Most starfish in shallow water crawl across the seabed mouth down, eating any small animals they come across or using their tube feet to overcome larger prey such as bivalves. *Freyella* belongs to a family of deepwater starfish, the brisingids, which feed upside-down, mouth upward, by extending their long, slender arms above the sediment. Their arms—they may have a dozen or more—bear rows of spines, so together they form a meshwork that can sieve out small crustaceans. These would eventually wriggle through the mesh, of course, but before they can do so they are caught by the fearsome pedicellaria, which project from the spines and act like muscle-powered pincers. Other starfish and sea urchins have pedicellaria that keep their bodies clear of intruders; crinoids don't have pedicellaria, which evolved after the split between crinoids and other echinoderms, and that is why they are bothered by crawling hordes of myzostomids. In brisingids like *Freyella* the pedicellaria have become modified as food-catching devices, and any copepod or shrimp they trap by closing on a projecting seta is inexorably frog-marched into the mouth as the arms bow inward, and swallowed.

Station 205: 13 November 1874

16°42′ N, 119°22′ E; South China Sea off Luzon, 1,050 fathoms

Challenger crossed the Sulu Sea without being troubled by pirates, and Nares began to thread his way through the shallow seas separating the islands of the Philippines. It was, with hindsight, the wrong choice: he would have found much deeper water off the east coast, where the Philippine Trench runs the length of Mindanao, but he might have been nervous of the onset of the monsoon. The ship paused briefly at Iloilo before steaming in light airs to Manila, which was sufficiently well known that Murray shrugs and says that it is "unnecessary to say much respecting it." They stayed there a week then left using both sail and steam as the weather began to break up, with variable winds and a heavy swell.

Two days later they were off the Bolinao peninsula at the edge of the South China Sea. In the morning the weather was fine enough to put the trawl over in a thousand fathoms or so, despite the heavy swell that continued to rock the ship. The trawl was recovered at 4:00 p.m. with plenty of animals together with a few pieces of wood and leaves. Wood had turned up in the trawl before, especially close to land, but showing that waterlogged wood will sink did not seem particularly interesting. It was not even edible, because lignocellulose is much more difficult to digest than leaves or flesh, which is why there are lots of dead trees on the forest floor but very few dead deer. Nevertheless, every sailor knew that there are some animals able to consume wood, because the wooden hulls of sailing ships, if left untreated, soon became riddled with the burrows excavated by their mortal enemy, the shipworm.

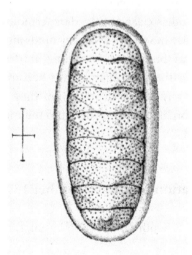

Figure 52 *Leptochiton* from Station 205

Source: A. C. Hadden (1886), Report on the Polyplacophora. Zoology, Part 43, Plate 1. Published in Reports, vol. 15.

Eating wood

The shipworm is not really a worm, although it is a long, soft, pinkish animal that looks like one. It is actually a bivalve like a clam whose shell valves have been reduced to two small curved sharp-edged triangles that it uses to cut a tunnel into logs that have been washed into the sea. The actual digestion of the wood is done, as in termites and other wood-eating animals, by microbes carried in the gut or some accessory organ. Shipworms relished wooden docks and piers, but their real breakthrough came with the invention of long-distance sailing ships in the fifteenth century, whose wooden hulls gave shipworms a world cruise with all they could eat. In this way, they quickly spread around the world, but always remained shallow-water animals. Far below the surface, however, was a hidden community of wood-eating animals whose existence was unknown until they began to be dredged up from the deep sea. They have evolved to eat wood because in the starving depths of the sea the arrival of anything organic, even a chunk of rotten wood, is a potential bonanza. Mollusks are well positioned to use wood because they can use their radula to scrape off the bacterial slime from the surface of a stick or log. Scraping will inevitably produce a cloud of sawdust that will be used by any microbes in the vicinity with the right mix of enzymes, and once they have found a way to stay with the animal that provides them with dinner they can be domesticated by their host, packaged and protected in special tissues, and put to work. This kind of symbiosis has evolved several times in deep-water snails and clams. It has also evolved in a less familiar group of mollusks, the chitons. Chitons have oval, flattened bodies and creep around on rocks, scraping off the algae, rather like snails, except that their bodies are protected by a series of calcareous bands rather than a shell. They are common in some places (such as the western coast of North America) and less familiar elsewhere, but are almost always found in the intertidal zone or just beneath; eating wood seems to be their only ticket to the deep sea. There is one small group of chitons found only in the deep sea that seems to have gone most of the way toward domesticating the full suite of wood-digesting microbes, but the animal trawled off Luzon, *Leptochiton*, has many close relatives in shallow water and seems to be in the early stages of acquiring the wood habit. It is not an impressive animal to look at—think of a small pale slug—but it certainly deserves A for effort.

The wind began to freshen an hour or so after the trawl was recovered and the monsoon broke in earnest that evening, with the main topgallant sheet carried away during a stormy night. The fresh northeast monsoon continued the following day until they reached the safety of Hong Kong harbor in the early afternoon of 16 November. They spent the next seven weeks in Hong Kong, and during this time the whole complexion of the expedition changed.

The change of command

While in Manila the officers had heard of an expedition to explore the Arctic coast of Canada, which at that time was still so poorly charted that an open Arctic Ocean was still thought possible. The frozen northern archipelago must have seemed very distant from the tropical Pacific, but it came much closer on 2 December, when Nares was handed a telegram from the Admiralty offering him command of the expedition. He seems to have demurred at first, wishing to complete the *Challenger* expedition before setting off on another, but a second telegram confirmed (and underlined) the appointment, and that was that: Nares was to be superseded as captain. It is not easy to say why. It might have been a perfectly normal appointment, of course, since Nares was an experienced surveying captain who had previous experience of the Arctic in the *Resolute*, searching for traces of the lost Franklin expedition. From this point of view he was a natural choice. On the other hand, their lordships may have viewed the progress of the voyage with some disquiet. The rate of desertion was truly shocking, and there had been several unfortunate incidents, such as the collision with the berg, which the ship had been lucky to survive. More men ran in Hong Kong, and the missing hands had to be made up with a draft of 60 men and boys sent out from London. The officers seem to have been fed up too, since half a dozen of them asked to return to England with Nares. These were not the symptoms of a happy ship, and perhaps their remote Lordships sensed this and decided to act at Hong Kong, the last big British naval base on the voyage where the transfer of command could be made in an orderly and routine fashion.

Wyville Thomson was greatly incensed and threatened to leave the ship and return with Nares, but was (rather easily, I think) talked out of this and stayed on. Balfour asked him to petition the Admiralty to allow him, Balfour, to return, so Wyville Thomson sent a telegram to Richards, "AS A FAVOUR GET BALFOUR APPOINTED TO ARCTIC STOP NARES PARTICULARLY WISHES IT STOP CAN BE SPARED FROM SHIP." With hindsight, the plea that a naval officer was superfluous may not have been the most persuasive way of putting the case.

Balfour waited anxiously for a reply, but none ever came, and when Nares left the ship, the only people to accompany him, besides his cabin steward, were Aldrich and the coxswain, both of whom he had specifically requested. There was a farewell dinner on 9 December that, according to Aldrich, "passed off very quietly," and the following day a boat party of junior officers—Bromley, Balfour, Channer, Hynes, Swire, and Havergal—pulled Nares across to the *Geelong*, which departed for England the next day. Aldrich was replaced by Lieutenant A. Carpenter from the *Iron Duke*. Nares was replaced by Captain Frank Turle Thomson, lately in command of HMS *Modeste*.

Thomson (always thus, to distinguish him from Wyville Thomson) came on board on 2 January 1875 with his fiddles, a piano, and a tomcat. The cat was to lead a hunted life: the men would surreptitiously take him up on deck at night to fight with the resident ship's cat, and he usually lost. As for Thomson himself, like any new captain he would have been preceded by his reputation, which would inevitably have fueled speculation among officers and men about how he was going to run the ship and disturb their settled ways. According to Matkin, Thomson "bears a bad name for tyranny on this station," which must have given plenty of scope for speculation, and he was also said to be fidgety and nervous at sea, a grave shortcoming. Neither of these misgivings lasted long; Thomson set out to be agreeable to the officers, and once at sea was confident and decisive in command. Come to think of it, Nares had been a bit of a fidget—Matkin even remembers him as "timid," which is strong language—although he seems to have been well liked, and was remembered fondly by everyone who recorded their impressions. Thomson was not physically imposing, being only five feet, four inches in height (162 cm) and weighing only 135 pounds (61 kg). Nevertheless, he came to be looked on as the stronger leader.

(The officers and scientists were weighed regularly during the voyage. Most were quite wiry: Nares and Campbell, for example, were only about 15 lbs. heavier than Thomson. Only Wyville Thomson scaled consistently above 200 lbs., although Chief Engineer Ferguson was almost as bulky; Moseley finished the voyage at 175 lbs. but for some reason shed 20 lbs. in the Pacific. The smallest officer was Spry, at 125 lbs.)

The transfer of command in Hong Kong was the hinge of the voyage, all the more clearly as *Challenger* now turned in her tracks and headed back toward the Philippines. The floors of the Pacific and South Atlantic were still to be explored, but most of the main scientific objectives of the expedition had been achieved by now: the deepest dredges yet attempted had shown that there were indeed animals living in the deep sea, far below any limit previously investigated, and many of them were new and strange. The indefatigable Murray had packed them all up (in 169 boxes) and sent them back to London in HMS *Adventure* on New Year's Day: money in the bank. Despite the vast reaches of the Pacific that lay ahead, *Challenger* was now heading for home, with little more than a year of the voyage still to come, and much of that in warm water. The change of captain, the turmoil of the last few weeks, the arrival of so many new faces, and, perhaps, the departure of the men most inclined to desert had all refreshed the expedition, and one might imagine (since there is no documentary evidence of it) a slight sense of elation at midday on 6 January 1875 as *Challenger* steamed out of Hong Kong.

14

Ninth Leg

The West Pacific Ocean, January–April 1875

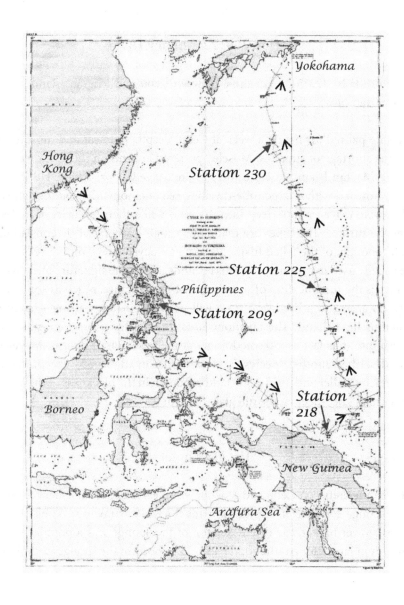

Full Fathom 5000. Graham Bell, Oxford University Press. © Graham Bell 2022.
DOI: 10.1093/oso/9780197541579.003.0015

Once out of Hong Kong *Challenger* made sail in a boisterous sea raised by the monsoon winds and headed back along her previous route, passing across the northern deepwater lobe of the South China Sea. Thomson's plan was to retrace the route as far as Zamboanga and then swing east far enough to make a fair wind of the northeast Trade to Japan. The first part of this plan was accomplished without much incident, apart from taking in tow an abandoned coaster and delivering her to Manila.

Station 209: 22 January 1895

10°14′ N, 123°54′ E; channel between Cebu and Mactan Island, 95 fathoms

The ship paused for nearly a week at Cebu, despite the heat and mosquitos and the shortage of supplies. Moseley pottered around on the muddy coral flats of Mactan Island, just offshore, where Magellan was killed. He found them crowded with a large tube-dwelling sea anemone, *Cerianthus*, whose tubes were driven a foot deep into the mud while their tentacles sprawled over the surface. The tubes are made of a tough leathery fabric consisting, oddly enough, of the felted filaments of nematocysts, the stinging cells that are usually used to capture prey. This reminds us that the signature of evolved change is the modification of some preexisting structure, much as someone with no access to a hardware store will patch their house using whatever materials lie at hand. The anemone uses its tube for protection and will vanish into it as fast as a burned finger at the first sign of danger. Moseley later described another species that was similar in all respects except that it was much smaller and lived 2,500 fathoms down in the blackness of the abyss rather than knee-deep in the full glare of a tropical sun.

The glass hotel

The channel between Cebu and Mactan was the site of a local fishery for the glass sponge *Euplectella*. Besides gracing Victorian drawing rooms as the Venus' flower basket, the dried skeleton of the sponge was a popular wedding gift in Japan, for a rather odd reason. The body of the sponge is a hollow

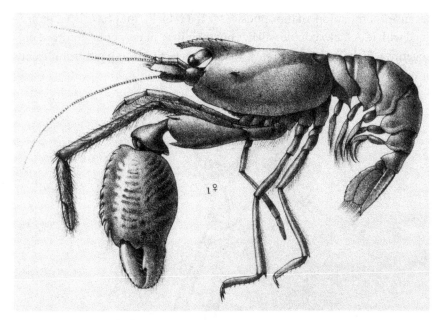

Figure 53 *Spongicola* from Station 209

Source: C. Spence Bate (1888), Report on the Crustacea Macrura. Zoology, Part 52, Plate 29. Bound in Reports, vol. 24.

cylinder of tissue supported by a meshwork of silica spicules, rather like a tower block whose walls are made of scaffolding covered with a sheet of canvas. It is quite well protected against being eaten because a bag of needles coated in toxic slime is not an appetizing prospect for a predator. The drawback of the strategy is that the hollow central core of the sponge makes a safe refuge for anything that can crawl inside, and sponges have no effective means of evicting squatters. Willemoes-Suhm noted at least four regular residents of the glass sponges trawled from this station: a polychaete worm, a small pale clam, an isopod, and a shrimp. The shrimp, *Spongicola*, is a specialist sponge-dweller (as its name implies) that is seldom found elsewhere. It enters the sponge as a newly hatched juvenile and grows up inside, secure against predators and (presumably) grazing on the plankton swept into the interior, like a guest in a hotel with a perpetual buffet and free accommodation. Once it is an adult, however, it has become too large to pass through the mesh of spicules that closes off the roof of the central cavity: its hotel has become its prison. It is not necessarily doomed to solitary confinement,

because a companion, of the opposite sex if it is lucky, may have been impris-
oned with it. A lucky couple will live out their lives in perpetual enforced mo-
nogamy, which is why the sponge was the perfect wedding gift from friends
or relatives who wanted to drop a hint.

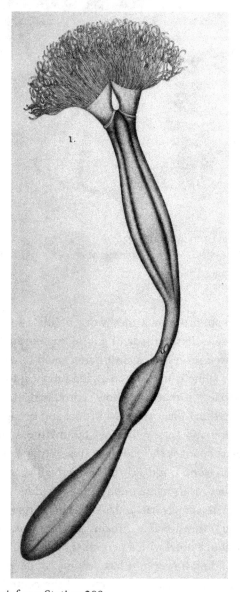

Figure 54 *Phoronis* from Station 209

Source: W. C. M'Intosh (1888), Report on the Phoronis Buskii. Zoology, Part 75, Plate 1. Bound in
Reports, vol. 27.

Metamorphosis

The ship trawled and dredged in shallow water off Malanipa Island, while a shore party explored the island itself, and found an abundance of life. Willemoes-Suhm noted in the Station-book that many of the animals quite literally lived on one another. There was a small snail studding the underside of holothuroids, a porcelain crab that clung to pennulataceans, a bivalve on a hydroid, and sipunculids living in the holes they had bored in the base of a coral. Some of these have little more affinity with their support than a bird perched on a twig, whereas others (like the snails) are deadly parasites. Somewhere in between is an odd wormlike animal called *Phoronis*, which the naturalists encountered here for the first and only time. It is shaped like a blunt cylinder, living in a tube that it secretes for itself. It has a crown of tentacles around the mouth and a long U-shaped gut. The tentacles are hollow and ciliated, forming a food-collecting device very similar to the lophophore of bryozoans and brachiopods, and used in the same way to sweep up bacteria and similar small prey. The intestine has to be recurved because a straight-through gut would empty at the bottom of the tube that the animal lives in, creating a health and safety problem. Other parts of the animal are quite sophisticated—it has proper blood vessels and even a simple kidney, for example, although its nervous system is rudimentary and it has no sense organs to speak of. The lophophore suggests a relationship with bryozoans and brachiopods, but otherwise it seems to be unique, a small, obscure animal blindly filtering the water of shallow seas, unconscious of its surroundings and known only to God and zoologists.

Its offspring grow up as larvae, and these were seen before the adult was known, because at certain times of year they swarm at the surface of the sea. It has to have a larva, because the body plan of the adult would not serve for a tiny animal, less than a millimeter long, which must feed itself as it grows up in the plankton. Starfish and crabs have larvae for the same reason. The larva is a minute, densely ciliated creature with tentacles and a straight gut, and this poses a problem: how does it make the transition between two completely different kinds of body? The first step is a prodigious heave that thrusts out its belly into a tubular shape, and at the same time bends its gut into a U-shaped curve. The whole of the forepart of the body, including the larval tentacles, is then sloughed off (and eaten), while the lophophore tentacles of the adult appear and begin to elongate. Having accomplished this startling metamorphosis in the space of 10 or 15 minutes, the young adult settles on the seabed and begins to secrete a tube for itself. At least, this is what it will do if there are no anemones around. If it happens to settle on one of the *Cerianthus* that live here, though, it somehow manages to insert its body into the column of the anemone, and spends the rest of its life as an uninvited lodger. There is no doubt that it stands to benefit from this arrangement, since the armed tentacles of the anemone will protect it from predators, while the anemone's gastric cavity—basically an open digestive vat—guarantees a rich local source of bacteria. The anemone gets no discernible benefit from the phoronid embedded in its body wall, but is unable to do much about it. A great many of these intimate associations between different kinds of animal are not cozy mutualisms from which both partners benefit, but simply freeloaders taking advantage of an unwilling or unconscious benefactor.

Thomson hoped to coal in Zamboanga, but after waiting four or five days in this "horridly tiresome place" (Campbell) decided to try Basilan Island instead, where he found a small stock of coal and set off eastward with enough to use in sounding and trawling. The ship passed north of the Talaud Islands, over the Philippine Trench and onto the abyssal plain at the southern end of the Philippine Sea. The wind grew light and variable and eventually failed entirely as the ship entered the vast abyssal plain of the West Caroline Basin, so that it had to proceed under steam, using up valuable coal. Passing over the New Guinea Trench, they approached the northern shore of New Guinea through a sea littered with driftwood so thick that the screw had to be stopped from time to time to avoid damage. The surface of the sea was alive with animals sheltering among the driftwood and the swarms of fish and small sharks feeding on them, and with the ship almost becalmed, Moseley could potter around in the cutter for hours adding to the collections. They anchored in Humboldt Bay on 23 February with hopes of exploring the almost unknown hinterland, but were met with implacable hostility by the local inhabitants and left, thoroughly frightened (in Campbell's words), after the first day. Matkin describes what happened when Nares set out with presents and the best of intentions: "Just as he approached the village a crowd of naked women rushed down to the beach, armed with spears, bows and arrows &c, & frightened our brave Captain on board his ship again in a very short time." Admittedly, officers and men were under very strict orders not to use firearms, but even so they must have reflected ruefully that a vessel of the Royal Navy had been put to flight by some angry women and a few tribesmen in canoes armed with bows and arrows.

Station 218: 1 March 1875

02°33′ S, 144°04′ E; Western Bismarck Sea, 1,070 fathoms

The following few days were fine, with light breezes, and the trawl was put over in a little more than a thousand fathoms as they entered the Bismarck Sea. It came up full of all sorts of animals. Among the least conspicuous (although one of the prizes of the expedition, according to Murray) was a small scallop called *Propeamussium*, about the size of a little fingernail, with thin, almost transparent valves that hid a remarkable body. Most bivalves are sedentary animals, either fixed immovably to a surface (like mussels) or able at best to shuffle slowly through the sediment (like clams). Scallops, however, can swim. They allow the valves to gape widely and then suddenly clap them shut, expelling a jet of water that forces them backward by reaction. *Propeamussium* has a band of strong muscle, directed obliquely, that can shut the valves very quickly; the uppermost valve has a flexible apron that increases the volume of water that can be expelled; and the closed valves have an airfoil shape for gliding. It is probably a good swimmer, at least by bivalve standards, although nobody has yet seen it swimming, or indeed doing anything else.

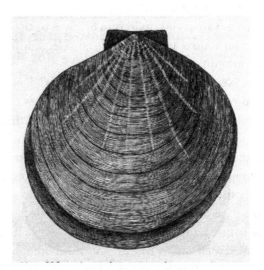

Figure 55 *Propeamussium* from Station 218

Source: E. A. Smith (1885), Report on Lamellibranchiata. Zoology, Part 35, Plate 22. Bound in Reports, vol. 13. Also in Narrative, vol. 1, Part 2 Figure 212.

A fierce scallop

Nobody has seen it eating, either, but it is certainly carnivorous, because its stomach is often crammed with larvae or small crustaceans. It captures its prey in much the same way as *Rhinoclama*, caught 18 months previously at Station 98 in the Atlantic: they are sucked in by a rush of water into the mantle cavity, then pushed into the mouth by the foot. Like *Rhinoclama*, too, it has a much smaller stomach than the capacious bag of filter-feeding bivalves. The structures that trap, sort, and process small particles, essential to most bivalves, are reduced in both species to the point where they no longer function, so both have become obligate carnivores unable to revert to their ancestral way of life. Despite their similarity, the two are not closely related, and must therefore have independently evolved the entire improbable suite of characters associated with eating small invertebrates. This is one reason that Darwin was so interested in insectivorous plants that he published a book about them. Plants that eat animals are even less likely than clams that eat animals, but nevertheless any good botanic garden has a greenhouse full of carnivorous plants belonging to half a dozen families. Bivalves lack their usual food, phytoplankton, in the deep sea; plants lack an essential nutrient, nitrate, in acid bogs. Capturing small crustaceans or insects supplies the want in both cases, and the habit evolves through the modification of some preexisting feature, such as valves and siphons in bivalves or the leaves of pitcher plants. Starting from unlike ancestors, natural selection tends to produce similar outcomes in similar circumstances.

Repulsed from New Guinea, *Challenger* finally sighted the Admiralty Islands in fog and rain on 3 March. It had taken 26 days from Zamboanga to reach the islands, so Thomson gave up the idea of pushing on eastward as far as the Carolines, and, after a brief stay, turned north instead, heading for Japan by the most direct route. As the ship steamed out of harbor there was one heart-stopping moment as she passed over an uncharted patch of coral with only about three feet to spare before the screw could be stopped. The ground must have been clearly visible from the deck. Once in deeper water she paused to trawl, and then trawled again in over a thousand fathoms just short of the West Melanesian Trench. This trawl brought up a rare deepwater octopus, *Japetella*, a fragile hand-sized animal with a translucent gelatinous body. In an adult female the ring of muscle around the mouth becomes transformed into a luminous photophore, so that she appears to have applied a smear of garish yellow lipstick. It is supposed to be a sexual signal, although this is little more than a plausible guess. The ship then followed, more or less, the northern spur of the trench, sounding in deep water of more than 2,500 fathoms but once again missing the bottom of the trench, which lay a thousand fathoms deeper still. She then sailed over a shattered sea floor studded with submarine hills rising to within 30 fathoms of the surface, trawling or dredging every two or three days. It was a weary time, reminiscent of the Ancient Mariner. The winds were light or variable, and from time to time the ship was completely becalmed. She plodded on at walking speed under the tropical sun, taking a month to cover the 2,600 miles from Manus Island to Yokohama. The days were variously described as wearisome, tedious, and monotonous. Any feeling of lightness and relief on leaving Hong Kong had been overtaken by exasperation and boredom. And yet it was during this leg of the voyage that two of the most important discoveries of the voyage were made.

Station 225: 23 March 1875

11°24′ N, 143°16′ E; Mariana Trench, 4,475 fathoms

The Challenger Deep

Early in the morning of 23 March the ship sounded in over 4,000 fathoms. After having sailed unwittingly over a succession of ocean trenches, the *Challenger* had finally succeeded, by pure luck, in dropping three cwt of metal (about 150 kg) attached to five miles of rope into the Mariana Trench. There was a little doubt, however, about exactly what had happened; Balfour, who was on deck at the time, gives the most complete account. The sounding apparatus was designed so that the weights were detached when the lead hit the bottom. The line could be checked from time to time to confirm, from the strain on the accumulators, whether the weights were still attached, which would show that the bottom had not yet been reached. The line had been checked at 3,000 fathoms (weights still on) and appeared to get no bottom at 4,575 fathoms, but the weights were off. When the line was hauled in the weights were indeed detached, but there was no more than a smear of clay in the sounding tube. It seemed likely that the line had been payed out faster than the weights took it down, so the true depth could have been anywhere between 3,000 and 4,575 fathoms. A second sounding was made to make sure, with four cwt on the line and frequent checks below 3,000 fathoms. This time, according to Balfour, "Bottom was got distinctly at 4450 [fathoms]." Murray says that the depth was between 4,475 and 4,500 fathoms. The three or four inches of sediment in the tube had a surface layer of dark brown clay packed with the siliceous skeletons of radiolarians, and paler layers of radiolarians and diatoms below. There was no other information: both thermometers were smashed by the pressure, and the pressure itself was beyond the range on the gauge. Records are often tendentious, but this was probably the deepest fully authenticated sounding yet made, and was certainly by far the deepest made by the expedition. It was not made at the deepest point in the trench, which lies about 50 miles to the west and is nearly 6,000 fathoms (35,800 feet or 10,900 m) deep. Curiously enough, it lies about the same distance below the surface of the sea as the summit of Mount Everest (29,000 feet or 8,840 m high) rises above it, although this seems to be merely a coincidence, like the similar apparent sizes of Sun and Moon. This deepest of pits, its position indicated rather than actually surveyed by the ship, is now known as the Challenger Deep.

Neither dredge nor trawl was deployed, although the dredge had been used two days before and the trawl was used two days afterward. Perhaps it was too late to use either once the second sounding had been completed, at 4:00 p.m., but it still seems that a great opportunity had been missed. If nothing else, a few animals from five miles down would have been striking emblems for the expedition, and the most concrete possible demonstration that Forbes's hypothesis of an azoic zone was fallacious. And yet: suppose that we do as Forbes did, and plot the number of species found in relation to depth, for all the stations sampled by *Challenger* during the whole of the expedition. Like Forbes's graph, this has a negative slope, with fewer species being found at greater depths. Like Forbes, we can extrapolate this graph to the depth where, on average, only a single species was found; below this lies an azoic zone in which most samples will be barren of life. I have drawn the graph and found this critical depth to be 5,400 fathoms (32,500 feet or 9,900 m), which is close enough, given the roughness of the estimates, to the bottom of the Mariana Trench. You do not need to take this observation seriously: the gear used by the *Challenger* was crude by modern standards, only large animals are included, and there are no samples much deeper than 3,000 fathoms. All very true; and yet the samples confirm Forbes's crucial observation, that the number of species declines at greater depths, and suggest that the edge of the azoic zone is at the limit of the sea. At the very least, traveling around the world with a dredge and identifying all the animals larger than a lentil that it catches gives a surprisingly accurate estimate of the greatest depth of the ocean.

Station 230: 5 April 1875

26°29′ N, 137°57′ E; near Susami Seamount, 2,425 fathoms

Challenger crossed the trench and continued parallel to and about 300 miles west of the Izu Trench, a southward extension of the Japan Trench, across a long oval abyssal plain that led to a complex series of troughs, ridges, and chains of seamounts. The light airs at least allowed soundings to be taken every couple of days, always in water nearly 2,500 fathoms deep, but the trawl yielded very little. It remained on deck, or came up almost empty, and on one occasion was lost when the line parted. The total haul for a thousand miles of sailing in calm seas was two small fish, three shrimp, and some fragments of a starfish. The shrimp were quite interesting; they belonged to a group of deep-sea shrimps that have the extraordinary habit, when attacked by a predator, of spitting out a great gob of luminous blue pigment. A sighted predator will be dazzled, or perhaps drawn to attack the glowing phantom, as the shrimp scuttles away into the darkness. Shrimps apart, this was a very meager total, given the opportunity of sampling a remote region of the deep sea. Perhaps the naturalists had lost heart as the ship dawdled on over the featureless ocean. But Buchanan had not, and as the ship approached Japan the chemist announced a startling and upsetting discovery.

Huxley's mistake

To understand what Buchanan was about, and why he startled Wyville Thomson and the others, we have to go back 10 years to Thomas Huxley's laboratory at the Royal College of Surgeons, where he was examining preserved samples of deep-sea sediment under the microscope. He noticed that the sediment was often glazed with a film of viscid material, lacking a distinct membrane but peppered with minute calcareous disks. This appeared to be some formless unbounded organism that Huxley named *Bathybius haeckeli,* after the eminent German zoologist and Darwinian. Haeckel was delighted by the compliment and soon found *Bathybius* for himself. Here was the missing link that connected animal with mineral, hidden in the deep sea: "Huge masses of naked, living protoplasm," wrote Haeckel, "cover the greater ocean depths." There was no doubt that *Bathybius* was widely distributed: Carpenter found it in sediment from the *Lightning* cruise, and when he dredged it from *Porcupine* in the Bay of Biscay it seemed to tremble under the microscope, further proof that it was truly alive. It fitted very neatly, too, with Dawson's *Eozoon,* which could now be interpreted as the fossilized remains of an ancient *Bathybius.* Everything, in fact, was coming together in a gratifyingly simple way: protoplasm is continually formed from inorganic material on the sea floor, where it extends as an almost complete organic sheet whose fossil remains are *Eozoon* and whose contemporary representative is *Bathybius.* Little more than a decade after the *Origin of Species,* a general theory of matter had emerged to explain the origin of life as well as the subsequent evolution of animals and plants from the formless protoplasm that anyone could see for themselves under the microscope. There was only one snag: the naturalists of the *Challenger* could not find *Bathybius.*

They had certainly tried. At the beginning of the voyage they had scrutinized every sediment core, without ever having that "Aha!" moment. Later on, they had all (except Murray) become rather bored with mud, but the continued absence of *Bathybius* was becoming a puzzle. Perhaps it was, after all, rather patchily distributed? Perhaps the coring technique was defective (not very likely) or Murray unobservant (even less likely)? The solution was eventually found by Buchanan, of all people, during the doldrums of the long haul toward Japan. He discovered that adding a moderate amount of alcohol to seawater resulted in a cloud of small crystals. The reason is that the alcohol reduces the solubility of inorganic salts, which crystallize out. The most common salt in seawater, other than sodium chloride, is calcium phosphate, which is what the crystals were made of. (The hydrated form of calcium phosphate is gypsum, used in making plasterboard. You probably have lots of it in your house.) More surprisingly, if he added a lot of alcohol the calcium phosphate left solution in the form of a gel, which looked something like protoplasm; sprinkled with a few of the calcareous disks common in marine sediments (they come from plankton growing near the surface), it looks very like *Bathybius.* It will even tremble a little if the ship moves, as ships do in the Bay of Biscay. Wyville Thomson was at first skeptical, perhaps in part because Buchanan was a little too inclined to crow about the triumph of simple chemistry over the airy speculations of the zoologists. But there was little doubt that Buchanan was right: *Bathybius* was an inorganic precipitate of calcium phosphate. Just as bad, King and Rowney were right, too: *Eozoon* was also inorganic, formed by the infiltration of magnesium salts. With its base removed, the elegant theoretical edifice built on *Bathybius* and *Eozoon* tumbled in ruin.

Once convinced, Wyville Thomson told Huxley of Buchanan's discovery that June. Huxley published part of the letter in *Nature* and promptly, and very creditably, abandoned *Bathybius*. Haeckel was more committed to his grand theory and continued to argue the toss for a decade; but after Huxley's recantation the notion of a universal sheet of simple protoplasm was dead, never to be revived. Except, of course, that it was, eventually. The area of the deep sea floor that is covered by the cryptic filaments of komokis is unknown, but in our present state of knowledge an estimate of 30 percent would not be unreasonable. Komokis are not unbounded protoplasm and they tell us nothing about the origin of life, but they do seem to provide a living frame for the uppermost layer of deep-sea sediment. They do not vindicate Huxley, but they do, perhaps, allow him a rueful smile.

Challenger sailed into harbor at Yokohama in the afternoon of 11 April and stayed in and around Japan for the next two months.

15

Tenth Leg

The North Pacific Ocean, May–July 1875

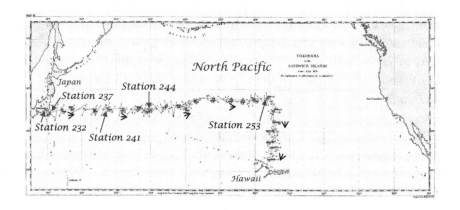

Station 232: 12 May 1875

35°11′ N, 139°28′ E; Sagami Bay, 345 fathoms

After a fortnight at Yokohama the ship was first dry-docked for repairs and then left for an extended cruise off the south coast and into the shallow sheltered waters of the Inland Sea. The naturalists visited a glass sponge fishery where rope sponges, *Hyalonema*, were caught. Their dried skeletons, long tubes of interlaced silica spicules, were used as ornaments, like the Venus' flower basket sponges they had seen in the Philippines four months previously. Long lines with barbless hooks snagged the rope sponges 300 fathoms deep and brought them to the surface along with a by-catch of pennatulids, echinoderms, and fish. The local gear proved to be more efficient than the dredge and trawl, which despite repeated attempts failed to catch any of the sponges (those reported from this station were presumably purchased from the fishermen). They did catch many other animals, including the giant spider crab, *Macrocheira*. This gigantic animal can measure well over three

Full Fathom 5000. Graham Bell, Oxford University Press. © Graham Bell 2022.
DOI: 10.1093/oso/9780197541579.003.0016

meters from claw-tip to claw-tip and is powerful enough to inflict painful injuries on an unwary fisherman. Like other spider crabs, it decorates its carapace with hydroids, sponges, and barnacles for camouflage; it might seem too large to need to hide from predators, but no doubt a big octopus could pull it to pieces.

Knots in slime

The dredge also contained an exceptional draft of fish, including deep-sea types such as grenadiers and cusk-eels as well as others that had probably been netted near the surface. Wriggling among them were half a dozen fish that looked like plump eels. These were hagfish, *Myxine*, and they were probably covered in slime. Hagfish produce great quantities of a superior kind of slime that expands on contact with seawater into hundreds of times its original volume. When a hagfish is bitten by a wrasse or a shark, it projects a jet of this slime into the mouth of its attacker, choking it or clogging its gills, so that it backs off, shaking its head and releasing its prey. Hagfish have no vertebrae or jaws: together with their near relatives, the lampreys, they are the distant descendants of the jawless fish that flourished 400 million years ago, long before the radiation of modern bony fish. It is not easy to bite without jaws, but they do their best. Their mouth is bordered by a horny plate that bears rows of sharp teeth. When they find a large dead animal on the sea floor they open their mouth wide, stab their teeth into the carcass, and then close it again. To remove a chunk of flesh, they can revolve rapidly to twist it out, or form an overhand knot with their tail, pass this knot the length of their body and then push against it to wrench out their meal. Or they can simply find a way in, through the mouth or anus or an open wound, and browse from the inside, which is why drowned bodies are often full of hagfish after a few days. They can also eat living prey such as small fish or shrimps, but they are not very skilled because their dental plate is a rather cumbersome apparatus that takes a long while—about a second—to operate. Jaws are much faster and more versatile, and once jaws had evolved (in a quite different way, through the gradual modification of the cartilaginous bars that supported the gills next to the mouth) the ancient jawless fish faded away, leaving only the hagfish to patrol the seabed, waiting for the next corpse to fall.

The ship trawled in the shallow coastal waters and also began to deploy deep tow-nets to sample the ocean above the seabed. They sounded in very deep water off Cape Shionomisaki, the southernmost point of Honshu, but could not trawl there because the wind was rising and the barometer falling. Their last attempt was on hard ground in Sagami Bay, where the trawl got stuck in the rocks and torn to pieces. They finally got back to the real business of the expedition on 16 June, when they left Yokohama flying the Homeward Bound pennant and sped on their way by "Home Sweet Home" from the ships of the Fleet.

Station 237: 17 June 1875

34°37′ N, 140°32′ E; Boso Canyon out of Yokohama, 1,875 fathoms

The ship rounded Nosima Point by night and set off to the southeast, sounding and trawling early in the morning about 40 miles offshore. The trawl came up in mid-afternoon stuffed with mud, stones, tree leaves, squid beaks, and fish bones, like a backstreet garbage collection, but also slithering with animals. There were gorgonians, starfish and brittle stars, isopods, brachiopods, a very deep crab, and many others. There were even two cephalopods. One was an inkless, deep-sea octopus; shallow-water species void a cloud of ink as a decoy when attacked, but in the lightless depths ink would serve no purpose. The other was a big-finned squid that remained the only representative of its species until a second specimen was captured nearby more than a century later. There was a good deal of interest later in an anemone that apparently lacked tentacles, having only blank openings around the mouth where the tentacles were expected. It was thought to represent the final stage of tentacle reduction in deep-sea anemones. In fact, *Liponema* is not sessile like most anemones, but instead rolls around on the sandy sea floor, shedding tentacles along the way. Each of the detached tentacles can then regenerate a complete animal.

The mop-headed animal

The most remarkable catch, however, was a hydroid. Many hydroids had been captured before, usually as inconspicuous felts on stones or shells formed by colonies of minute polyps. They are very beautiful under the microscope, like water brought alive and made visible, but in the air they collapse and can

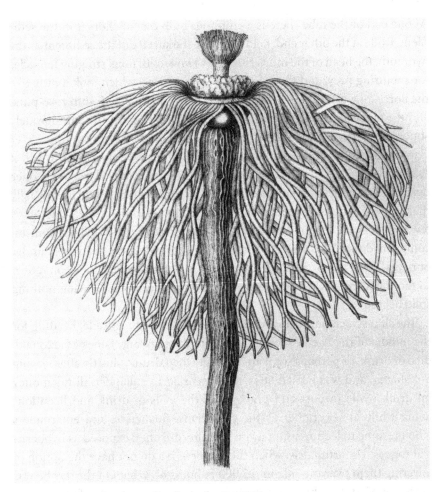

Figure 56 *Branchiocerianthus* from Station 237
Source: G. J. Allman (1888), Report on Hydroida. Second, Part. Zoology, Part 70, Plate 3. Bound in Reports, vol. 23.

easily be passed by as little more than a smudge. This hydroid was not likely to be passed by because it was a gigantic solitary polyp over seven feet tall, taller than Wyville Thomson, albeit much slimmer. Draped across the trawl, it looked like a long-handled mop, or, better, a length of garden hose ending in a mop head. The hose part is a hollow tube less than an inch wide with a single layer of cells on the outside and a single layer on the inside, separated by a firm but very extensible jelly; the specimen in the trawl was probably fully stretched out and might be much shorter in its normal attitude.

At one end of the tube there is a chitinous bulb that anchors it in the sediment, while at the other end, held far above the surface of the sediment, is the hydranth, the head of the mop. This bears two arcs of long, stinging tentacles for capturing prey, and the reproductive structures, which look a little like the florets of a cauliflower. The stalk was pale pink, the hydranth rose-pink, and the tentacles scarlet. This gorgeous animal was given the appropriately sonorous name of *Monocaulus imperator*, although the taxonomists later claimed that *Monocaulus* had been used before, so we now have to call it *Branchiocerianthus*, which sounds merely fussy. Its beauty is transient; once dead, hydroids deliquesce. They can be stored in spirits, but this usually makes them contract violently and tear themselves to pieces, leaving only a few rags of tissue for the zoologists to study. A frustrated Wyville Thomson complained that "it is wretched to see them melting away absolutely under our eyes." Even the largest hydrozoans, the water animals, all too soon return to the water from which they seem to have been sculpted, leaving nothing solid behind.

The ship struck out eastward along the 35° N line of latitude, heading for the middle of the Pacific Ocean. A sick-berth attendant, James Macdonald, died of drink, or perhaps from an accident when drunk, shortly after leaving Yokohama, and was buried at sea. He was said to squander all his money on drink while in port and prate about the evils of drink and licentious living while at sea, rather as the politicians, financiers, and entertainers who presume to lecture us today on morality are often exposed as hypocritical rogues. Unfortunately, when this happens we do not have the option of lowering them over the side into 3,000 fathoms of water. On the day he was buried, indeed, the ship sounded in very deep water of 3,950 fathoms, the second deepest of the cruise, near where the American survey ship *Tuscarora* had reported 4,655 fathoms the previous year. I cannot account for it. There is even deeper water at the northern end of the Izu Trench, but *Challenger* must have crossed this (as usual without noticing it) the previous night, and unless the navigation erred by two degrees of longitude (inconceivable), she was in water 1,000 fathoms less deep. The following day she sounded in 3,650 fathoms, which seems equally unlikely. She was at this point sailing into the great pond of the central Pacific Ocean, ringed by trenches that trap the sediment washed from the land. There are in any case few rivers that match the output of the Congo, Mississippi, and Amazon, so the sediments tend to be thinner and the topography more rugged than the Atlantic. The water is certainly deep, generally between 2,500 and 3,000 fathoms, but seldom much more; perhaps the sounding crew had become a little rusty during their long stay in Japan.

The ship sailed on over this vast abyssal plain for the remaining 3,000 miles to Honolulu. Forbes's dream had come true. Forty years before, he had pulled an oyster dredge over the seabed by hand in 20-fathom water off the coast of Isle of Man; now, a naval ship with a crew of 200 could use the latest machinery to wind in four miles of rope and explore the bottom of the largest and deepest ocean in the world. For all that, there was a lackadaisical air on this leg of the voyage. The wind was fair, although often light and variable, but the skies were generally overcast and drizzly, and the days seem to have passed by slowly. Campbell and Moseley ignore this spell completely; "very little of interest occurred from day to day," according to Spry; and even the meticulous Murray, who gives a day-by-day account of most of the voyage, skims over this leg. Matkin was in low spirits because news of the death of his father had reached him in Japan. Even Willemoes-Suhm was feeling more and more homesick.

Every three or four days the naturalists attempted to dredge, but they were not very successful. The trawl carried away twice, losing not only all the apparatus but also some 50,000 feet of rope, which must have depleted even the vast stores carried by the ship. The rope fouled the screw on one occasion, and the trawl had to be recovered prematurely on another when the weather grew threatening. When the trawl could be deployed and recovered it was often torn by the rocky ground and retained only a few animals. It was not all that Forbes might have hoped for, although it did give a final answer, in Murray's opinion at least, to the question that Forbes had originally raised.

Station 241: 23 June 1875

35°41′ N, 157°42′ E; northwest Pacific Basin, 2,300 fathoms

The sea became noticeably colder and took on an odd appearance, with areas of red and white, so striking that Thomson had the hand lead cast in case they were shoals. The tow-net showed that the patches were immense congregations of copepods, the red alive and the white dead—perhaps killed by cold water from the north. The trawl was put over, but the rope parted as it was being recovered. The next day they tried again, but while the rope was being payed out it fouled the screw and the trawl was lost. Finally, after

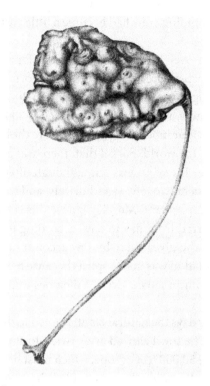

Figure 57 *Culeolus* from Station 241

Source: W. A. Herdman (1882), Report on Tunicata. First, Part. Ascidiae Simplices. Zoology, Part 17, Plate 10. Bound in Reports, vol. 6.

these two unsuccessful attempts, the trawl brought up a good haul of animals on 23 June. Some of these were old acquaintances, members of the deep-sea fauna with which they were becoming familiar. There were glass sponges and pennatulids, a cup coral, deep-sea ophiuroids and holothuroids that they had seen before, the wood-eating chiton, and an assortment of annelids and crustaceans. There were also bucketfuls of pumice stones, which are thickly strewn over the seabed in this region, offering hard standing for sessile organisms and a very convenient sampling device for naturalists.

The bamboo grove

One of the animals attached to the pumice was the bamboo coral *Bathygorgia*, a relative of the black whip coral caught at the start of the voyage off Cape St.

Vincent. It has pronounced nodes separated by white calcareous internodes and looks superficially like a stem of bamboo. The polyps are spaced along the stem, which serves to raise their tentacles above the sediment. There was also a rather drab tunicate, *Culeolus*, that they had seen a couple of times before from deep water. Most tunicates, like the *Abyssascidia* dredged the previous year on the way to Australia, are filter feeders that suck water in through one siphon, strain out the phytoplankton through slits in the pharynx, and expel the effluent through a second siphon. The stomachs of deepwater tunicates like *Culeolus*, however, contain copepods, amphipods, and polychetes: they are carnivores. Presumably they capture their prey by an exaggerated version of normal suction, triggered by the proximity of prey, but the mechanism has not yet been worked out. More importantly, tunicates are the third group of inoffensive filter feeders to have evolved carnivorous habits in the deep sea, joining sponges and bivalves. The Bellman Theorem ("What I tell you three times is true") suggests that this is a consistent trend, without explaining why. In shallow, sunlit seas there is an abundant supply of phytoplankton for animals like these to consume. In deep water, filter feeders must rely on dead cells sinking from the surface, and, since these have been asset-stripped by fungi and bacteria on the way down, they have a strong incentive to change their ways. This is a conventional explanation and seems plausible enough to serve for the time being.

Station 244: 28 June 1875

35°22′ N, 169°53′ E; northwest Pacific Basin, near Emperor Seamount Chain, 2,900 fathoms

The ship moved on at walking pace. The trawl was put over three days later, but the rope broke at a block, so the trawl and all 15,000 feet of line were lost. The next time they were more successful; the net came up torn but full of mud, pumice, nodules, shark teeth, and animals. Teeth are the hardest part of a shark, and those from a big shark are massive enough to last for ages in the mud. Some become immortalized by forming the nucleus of a nodule, which, like a pearl, needs some seed around which to accrete. Other nodules form around pumice stones, whale vertebrae, and glass sponges. There are a lot of nodules—one haul yielded a bushel (about 36 liters) of them, which spilled from the net and bounced on the deck like a sackful of potatoes. There might therefore be a lot of large dead sharks falling to the bottom of the sea, which starts a chain of thought I shall continue later, at Station 286.

Diversity at depth

There were a lot of animals, encouraging Murray to be triumphal: "This was the most successful haul hitherto made from a depth of 2,900 fathoms, and proved the abundance of highly-organised life in great depths." One might think that the point had been amply proven during the last two and a half years, but Murray can be forgiven for adding the underlining. Most of the animals belonged to what they could now recognize as the typical fauna of the abyss: a range of soft corals, the cup coral *Bathyactis*, a crinoid, a brachiopod, a couple of annelids, and the usual collection of echinoderms, including starfish such as *Freyella* and the deep-sea echinoid *Pourtalesia*. There were also half a dozen bivalves; these were the deepest bivalves found on the voyage apart from a solitary animal captured later in the month. One of the annelids was a minor mystery. It was a common sort of worm that secretes a calcareous tube to live in—you can find its white scribbling on pebbles in rockpools almost anywhere. At this depth, however, calcium carbonate should dissolve, so how it had managed to make a tube is a puzzle. Otherwise most of the animals were familiar, although some were different enough from anything yet found to be described as new species, three of which were named after Wyville Thomson: a starfish, a holothuroid, and the brachiopod. Murray bagged one of the bivalves.

Small fry

Four days later, trawling at about 2,000 fathoms brought up even more animals, although with hindsight it is clear that many were being lost. With steam in the boilers, the donkey engine would recover the dredge from 2,000 fathoms in about three hours. As it rose through the sea at a rate of about a foot per second (rather slower than the average escalator), it would trail behind a thick plume as the mud billowed from its mouth and sifted through its mesh. During the whole of this time small animals were spilled or winnowed through the mesh of the net, and by the time that it was dumped onto the dredging platform it would have lost most of the very small animals it might have captured on the bottom (and added others that it had come across in midwater). Moreover, as the trawl was being dragged along the seabed, any active animals could have scuttled out of the way, or were deflected by the pressure wave. It is not surprising that large sedentary animals like gorgonians and bryozoans and echinoderms usually filled the net. They were easier to find, too, if you imagine yourself groping through a small mountain of mud, on a sailing ship, in a moderate sea. Animals a couple of millimeters long would be likely to escape notice, even if they remained in the net after its long journey to the surface. Take nematodes, for example, the small javelin-shaped roundworms that are so abundant in soil that a cupful may contain hundreds of individuals. We now know that small nematodes are also abundant in deep-sea sediments, but not a single one was recorded from the expedition. No doubt most other small fry were also lost or overlooked, but on this occasion a couple of small crustaceans remained in the net and were picked out by the naturalists.

One was a cumacean, dredged up from the shallow burrow it had scraped in the top of the sediment. Cumaceans look like skinny shrimps wearing an enormous hood. The hood is an inflated carapace that covers their head and thorax, enclosing a spacious chamber through which a stream of water is pumped by the beat of their appendages. Long, closely spaced hairs on their mouthparts filter out any edible particles, which are passed to the mouth and swallowed. Others feed by scraping the film from sand grains, tossing the cleaned grain aside like a cherry pit. The second crustacean was a tanaid, which looks quite different—something like a stretched wood louse—but seems to lead a similar way of life, living in a burrow and filtering detritus

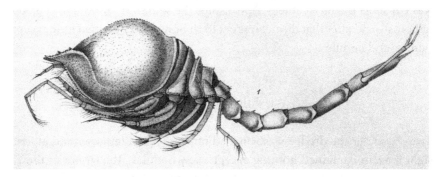

Figure 58 *Diastylis* from Station 244
Source: G. O. Sars (1887), Report on Cumacea. Zoology, Part 55, Plate 10. Bound in Reports, vol. 19.

or scraping sand grains, although some species may eat very small animals (such as nematodes). Both cumaceans and tanaids are minute, at most a few millimeters in length, and very few were collected by the expedition. Modern surveys using more sensitive sampling equipment have shown that both groups are much more abundant and diverse in the deep sea than the *Challenger* naturalists could have suspected. They are probably very important in processing every edible speck that falls onto the sea floor, like ants in a forest, but despite the modern surveys their ecology is still poorly understood.

Station 253: 14 July 1875

38°09′ N, 156°25′ W; Mendocino Fracture Zone, 3,125 fathoms

At noon the following day *Challenger* crossed the date line at 180 degrees of longitude. Unlike the Equator, this is a purely artificial line, and it has never been celebrated with the same humiliating ceremonies, but remains obscurely unsettling. It may be illogical to ask what happened during the day lost when traveling from east to west, or to wonder what to do with the day gained when going from west to east, but everyone does anyway. Thomson solved the problem by decreeing that the next day, Sunday, 4 July, should be repeated the following day, to give the crew a week with two Sundays. This was no doubt a popular decision, as Sunday gave a welcome break to the usual routine of the ship. The following day—I mean the day after the double Sunday—the trawl came up torn from 2,900 fathoms but contained a few animals, including a second gigantic *Monocaulus*. The next attempt to trawl was cut short by the weather—a pity, since the sounding showed over 3,000 fathoms—and after that the net was cut to ribbons and contained little except pumice and nodules.

An enigmatic polyp

Two days later the dredge was winched up from 3,125 fathoms, and at first sight it again contained nothing except a few nodules. The largest of these, however, was a foot long and had been colonized by a variety of animals. None of them were out of the way: komoki tubes, some bryozoans and hydroids and a small tunicate on the upper surface and annelids underneath. They were undoubtedly from the bottom, of course, and turned out to be the deepest animals collected during the expedition; a pity they were not more distinctive. A few small tubes belonging to an animal called *Stephanoscyphus* were also mentioned in the Station-book, as an afterthought. This was not much out of the way, either, turning up in most of the hauls from the Pacific, but what sort of animal was it exactly?

The previous year, while the *Challenger* was dredging in Sydney harbor, a well-known naturalist called George Allman was paddling in the shallows at Antibes (I think he must have been on holiday there). Being a naturalist, he was looking at the sponges growing there, and being a good naturalist he noticed that one looked odd—the pores in its body wall seemed unusually prominent—so he cut off a piece, put it into a vial, and looked at it through a lens. What he saw, to his astonishment, was a crown of tentacles unfolding from each pore. At first he thought he had discovered the link between sponges and hydroids that, in the heady decades following the publication of the *Origin of Species*, some biologists believed to exist. When he examined the sponge more carefully, he soon discovered his error, but the truth was just as surprising. The tentacles did not belong to the sponge at all; they belonged instead to a complex branching colony of hydroids whose tubes ramified throughout the body of the sponge and whose tentacles poked out of its orifices. It was just like the hydra-headed worm that had been found inside sponges the previous September, off the Aru Islands, with each section of its body lodged in one of the rooms provided by its host. Allman labeled his find *Stephanoscyphus* and enrolled it as the founding member of a new category of hydroids, distinguished by its habit of secreting a chitinous tube. Very similar tubes with similar occupants were found afterward growing singly on hard surfaces in deep water, including the big slab of manganese brought up here. Now, George Allman was an expert on hydroids—he wrote the two voluminous Reports on the hydroids collected by the *Challenger*—but he was wrong about *Stephanoscyphus*, which is not a hydroid at all.

If he could have cultured the animal and followed its development, he would have seen that sooner or later it would divide transversely to form a column of disks like a stack of biscuits. Each disk would then be released, starting at the top, and float away as an independent animal shaped like a minute plate with tentacles around its margin. This animated asterisk is called an ephyra. It grows in size by eating the small animals caught by its tentacles and eventually develops into a bell-shaped medusa that is instantly recognizable as a jellyfish. The medusa is a sexual animal: it produces eggs or sperm, and the fertilized egg develops into a larva called a planula that looks like an elongate furry football and swims around for a time before settling onto a hard surface and growing into a polyp, which then repeats the whole complicated process. Whether polyp and medusa are two different individuals or two forms of the same individual is an intriguing question. If the polyp committed a crime, could the medusa be charged? Setting this aside for the philosophy class, there is a large and diverse group of animals, the medusozoans, whose life cycle includes a medusa. Their sister group is the anthozoans, including corals, sea anemones, gorgonians, and pennatulids, which live exclusively as polyps, or colonies of polyps. The two most familiar kinds of medusozoan are the hydrozoans, or hydroids, and the scyphozoans, or jellyfish. The production of ephyrae marks *Stephanoscyphus* as a scyphozoan; the animal found in sponges by Allman eventually grows into a common jellyfish called *Nausithoë*, but he was not to know. Polyp and medusa are as unlike as caterpillar and butterfly, and are usually given different names when first described, leading to a nomenclatural tangle that taxonomists spent lifetimes trying to straighten out.

At this point in the voyage the Captain changed his mind. The original intention had been to sail across the Pacific from Japan to Vancouver Island, then pass southward down its eastern margin toward Cape Horn. Thomson decided to abandon the rest of the route to the west coast of Canada and instead to turn south immediately. The ostensible reason was shortage of time; but I imagine that the prospect of rest and relaxation in Honolulu, only a thousand miles away, was also a consideration, after a tedious month that seemed at the time to have been spent mostly in losing gear or hauling in ripped nets containing little but pebbles.

16

Eleventh Leg

The Length of the Pacific Ocean, July–November 1875

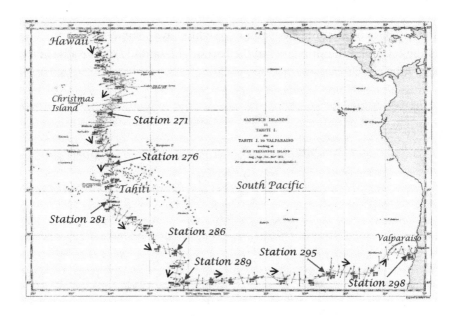

The change of course did not bring about a change of luck; quite the contrary. All the way to Honolulu there was only one successful dredge, and that produced only a single animal, a small bivalve that was undistinguished in every way except that it turned out to be the deepest bivalve dredged up during the cruise. At other stations the rope broke, or the dredge came up fouled, or only a sounding was taken, and the crew was becoming bored and restive. Campbell is clearly fed up with the cruise by now:

> Our work in that way is to us naval men, who have to work the subject practically, exceedingly wearisome. . . . And it is no good arguing that this work

Full Fathom 5000. Graham Bell, Oxford University Press. © Graham Bell 2022.
DOI: 10.1093/oso/9780197541579.003.0017

is our raison d'etre, we know that too well, and obey the law of our being to perfection. But we should be more than mortal if after more than two years of the "same old grind" we did not—but bah! Why growl? A splendid ship, a splendid cruise, romance, "how interesting!" &c., &c.—by all means, be it so. Has not Huxley said that our work in the Atlantic alone more than re-paid all the bother and expense of outfit? How fervently did we agree with him, and wished that the authorities had thought so too, and ordered us home to do no more!

After 10 days of this routine Oahu was sighted, with some relief, and after taking on a pilot the ship entered the harbor at Honolulu on the afternoon of 27 July.

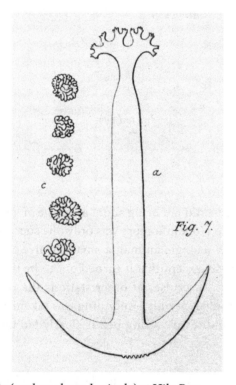

Figure 59 *Synapta* (anchor-shaped spicule) at Hilo Bay

Source: H. Théel (1886), Report on Holothuroidea. Second, Part. Zoology, Part 39, Plate 1. Bound in Reports, vol. 14.

The Admiralty worm

Challenger stayed in the Sandwich Islands, as they were then called, for three weeks, shifting to Hilo Bay for the last week. The group was later annexed by the United States as Hawaii and is familiar for surfing and the active volcanic crater at Kilauea. It is not surprising that the officers and naturalists should have visited Kilauea, which reminded them of the furnaces seen by night in the Black Country. It is a little more surprising that Campbell, who had evidently recovered from his ennui, took out a board and went surfing. He thoroughly enjoyed it, too, although when one of the men on the Transit of Venus team tried it later, he overturned and was drowned. Moseley went out dredging in the steam pinnace and collected the usual shallow-water fauna of corals, bivalves, snails, bryozoans, and reef fish. The only other zoological content was provided unexpectedly by a visit to the ship by Kalakaua, then King of the Sandwich Islands. He was shown a preparation of the skin of the holothurian *Synapta* under the microscope—it seems an odd way to entertain royalty, but there you are—and to everyone's surprise recognized it immediately. *Synapta* is a very elongate sea cucumber that looks a little like a rather plump snake. It has abandoned the tube feet characteristic of other echinoderms in favor of minute calcareous ossicles shaped like anchors with an oval base plate; perhaps they provide traction when burrowing, and they certainly make the animal stick to cloth, even to wetsuits. It is quite common in shallow water and doubtless familiar to the islanders, but even so this was perhaps the only occasion on which royalty has ever displayed an intimate knowledge of holothurians. Even the journalists were impressed, and from the day of the King's visit referred to *Synapta* as "the Admiralty worm."

Station 271: 6 September 1875

0°33′ S, 151°34′ W; central Pacific Basin, 2,425 fathoms

The expedition left Hawaii on 19 August and continued south over very deep water floored with the inevitable red clay. It is a difficult passage for a sailing ship, and *Challenger* was close-hauled most of the way because of unfavorable winds and current. The general air of grumpiness returned as the ship mooched along, accomplishing very little. The trawl did not reach bottom; then it was torn, then it came up intact but containing only a single mutilated

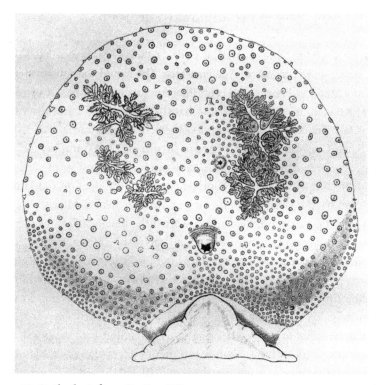

Figure 60 *Bathydoris* from Station 272

Source: R. Berg (1884), Report on Nudibranchiata. Zoology, Part 26, Plate 12. Bound in Reports, vol. 10. Also in Narrative, vol. 1, Part 2 Figure 238.

small fish, then it was brought up foul. Murray and Buchanan amused themselves by inventing a simple device to measure the transparency of seawater, consisting of a white plate a foot square that is lowered on a rope until it is no longer visible from the surface. Nowadays a modified version is called a Secchi disc and is a familiar tool of limnologists. The Secchi depth of an average lake might be about 10 feet or a little more, but in the clear water of the central Pacific, Murray and Buchanan could see their plate 150 feet down.

The beauty of the deep

Their luck turned on 6 September, when the trawl finally came up from deep water full of animals. There was a small fish with a big head and a mouth to match, belonging to a family of deep-sea fish, the "pricklefishes," that may be quite abundant but are still almost unknown; a century later this was still one of only four known specimens. A nudibranch that was later named *Bathydoris*, meaning "the beauty of the deep," was a more obvious surprise. Nudibranchs are marine mollusks that lack a shell and look rather like slugs except that they are often decked out in brilliant colors, probably to warn predators not to eat them. They are large attractive animals that are easily collected in shallow water, especially in the tropics, and naturalists have always found them irresistible. They don't belong in the deep sea, however, and this single animal was the only specimen that the expedition found at abyssal depths. It has the transparent, gelatinous body of so many deep-sea animals, but it is as showy as its shallow-water progenitors: orange gills and external genitalia, a dark purple foot, and brown gills. None of this can be seen, of course, in the perpetual darkness of the deep sea. Thomas Gray wondered why:

> Full many a gem of purest ray serene
> The dark unfathomed caves of ocean bear.

And he might have been equally puzzled by the futile splendor of *Bathydoris*.

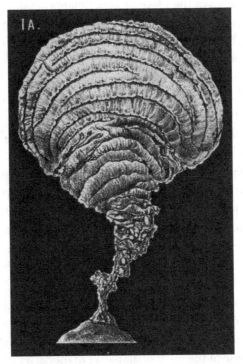

Figure 61 *Stannophyllum* from Station 271

Source: E. Haeckel (1889), Report on Deep-Sea Keratosa. Zoology, Part 82, Plate 1. Bound in Reports, vol. 32.

More strange lumps

The most abundant creatures in the haul were large xenos that Haeckel (thinking they were sponges) named *Stannophyllum*. They are more graceful than the usual xeno, which all too often is mangled by the dredge into a mere lump, because their body is supported by cables made up of very fine fibers and can develop as a thin, flexible, leaflike structure attached to some hard substrate (usually a nodule) by a narrow stalk. They are quite large—as big as a man's hand—and to some resemble a slender forearm holding a fan, although a more prosaic image is that they look like bracket fungi. There were hundreds of them in the net, more than had ever been collected before, and they are probably very abundant organisms. There are certainly large areas of the deep sea floor where every square foot has a xeno, and one-half or more of the total mass of organisms is made up of xenos. It seems inevitable that they play a proportionately large part in the ecology of the deep sea. The base of the food chain on land is made up of green plants that knit carbon dioxide from the air into sugars that can be eaten by animals, and at the surface of the sea the place of green plants is taken by single-celled phytoplankton. Neither plants nor phytoplankton can live on the deep sea floor, where the base of the food chain is instead provided by xenos and komokis, giant protozoans that transform the thin rain of particles from the surface into biomass that can be consumed by animals such as isopods and holothurians. A scheme of this sort might be worked up into a general theory that begins to make sense of how nutrients are circulated in the economy of the deep sea. There are a few remaining difficulties, however. One is that living cytoplasm is only a small fraction, probably only a few percent, of the total mass of these creatures, most of which consists of the debris they stick together to cover themselves or the peculiar rounded masses, possibly fecal pellets, that occupy much of their interior. It does not help that we do not really know what xenos and komokis eat, although there is no shortage of guesses. We do not know what eats them, either; all we have is a series of interesting anecdotes. A general theory will have to wait until these details have been cleared up.

Xenos have also suggested a very different general theory (ignorance always provides fertile ground for speculation) about the nature of the mysterious large fossils found at Mistaken Point that predate the radiation of animals. Adolf Seilacher, one of the most imaginative of modern paleontologists, has asserted that these creatures were constructed as thin quilts, subdivided into compartments, which might lay flat or be raised above the surface by a stalk. He rejects the usual notion that they are the earliest ancestors of modern animal groups, and interprets them instead as giant protozoans, similar to modern xenos, that evolved large size before animals and by a quite different process of development. If he is right, the contemporary xenos found in such profusion at Station 271 belong to a lineage that reaches far back in time, beyond the base of the Cambrian and the origin of animals, and are the most ancient inhabitants of the sea floor. It is only fair, although deflating, to add that few other biologists share this grand vision and most prefer instead to believe that xenos diverged from a foram stock in relatively recent times.

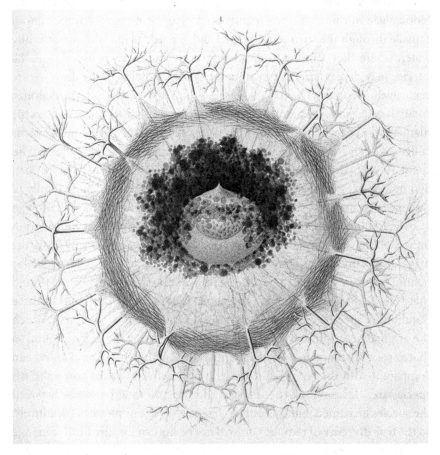

Figure 62 *Auloceros* from Station 271

Source: E. Haeckel (1887), Report on Radiolaria. Zoology, Part 40, Plate 102. Bound in Reports, vol. 18.

Glass mines

The other creatures that were extraordinarily abundant and diverse at this station were radiolarians, single-celled predators that are big enough to be seen with the naked eye. Under the microscope their skeletons of opaline silica are stunningly beautiful, forming helmets or bells or spheres, or spheres within spheres, pierced by pores and decorated with spines, endless combinations of form like illustrations of some fairyland geometry. Their cells have two compartments, like the traditional hut of medieval peasants with one room for the family and another for the livestock. The central compartment houses the chromosomes and the rest of the genetic machinery, separated by a wall from an outer layer of cytoplasm where the livestock are kept—symbiotic algae in this case. Fine filaments of cytoplasm pass from the central capsule through the surface of the cell and extend far into the surrounding water, where they wrap around passing prey. Radiolarians living near the surface may depend on their symbionts, but those living in the deep sea are exclusively predators, like tiny glass mines that capture the bacteria, diatoms, small larvae, and even copepods that bump into their long contractile filaments and digest them in the outer layer of cytoplasm. All the radiolarians collected during the voyage were later passed to Ernst Haeckel in Jena, who spent 10 years describing them—over 4,000 species in all, with more than 500 new species from this station alone (he even named a whole family of radiolarians after the ship—the Challengeridae). His report, weighing in at over 1,800 pages plus illustrations, was the main source of information on radiolarian diversity for the next 100 years, although for many years the original specimens it was based on could not be examined because Jena lay behind the Iron Curtain. After the reunification of Germany, it was found that Haeckel had been rather casual about keeping type specimens of these hundreds of species (a type specimen is an example of the material on which the original description was based, permanently preserved in a museum, so that anyone wishing to check the identity of something they have found can compare it with the type). This makes it impossible to know how valid his species are. Haeckel used the shape of the skeleton to differentiate between the species he named, but this might change with age or physical conditions, so the true diversity of radiolarians will not be known for sure until someone undertakes the mammoth task of reviewing the group with the same assiduity as Haeckel.

There is actually rather a lot that we don't know about radiolarians. They probably do not even form a natural group, but rather a motley crew of organisms with similar kinds of glassy skeletons. All are extremely difficult to maintain in the laboratory, and so far nobody has been able to follow a single species through its entire life cycle. They may reproduce by binary fission, but their elaborate glass skeleton cannot divide, and it is a puzzle to know what happens to it; sometimes the central capsule divides into lots of tiny motile cells that might be able to grow into the adult radiolarian, but in culture they always die. Alternatively, these motile cells might be gametes, but no sexual phase has yet been identified. Consequently, they cannot be crossed to investigate heritability—whether the form of their skeleton is heritable, for example. Come to that, it is not really clear how the skeleton is formed or what function it serves. One large group, the acantharians, makes its skeleton from strontium rather than silicon; there is plenty of strontium in seawater (strontium and silicon have similar concentrations), but no other animal needs it or uses it, and why acantharians should be so idiosyncratic is unknown. Radiolarians have been found to eat a range of other organisms, but how choosy they are is not clear; salps eat them, but how important they are as food for other animals is difficult to say. What is known for certain is that they are extremely abundant: there are often several hundred per cubic meter, or about a dozen in an average bucketful of seawater, or, if you prefer, about as many radiolarians in the world ocean as there are stars in the universe. Being so numerous, they are probably very important in the food web of the sea; it is a pity that, more than a century after the radiolarian bonanza at Station 271, we know so little about them.

The death of a naturalist

This station was a welcome relief from the slim pickings they had become used to, and in fact turned out to be the richest haul they took between Yokohama and Valparaiso. Even so, sorting the spoil was done under a shadow. The reason sounds almost comical: Willemoes-Suhm suffered from boils. Boils may be painful, but they are not usually thought of as being dangerous—undignified, perhaps—but they are nevertheless streptococcal infections that turn deadly if they find a way into the bloodstream. John Stuart Mill died of boils, or more precisely of erysipelas, a disfiguring infection of the skin, in the year that *Challenger* sailed. Willemoes-Suhm had a small boil on his face while they were in Honolulu, although it does not seem to have impaired his social life, since he seems to have got on very well with a sister of the King. He left Hawaii in good spirits and looking forward to the homeward leg: "After doubling Cape Horn, we head home, God be thanked!" he wrote to his mother. A few days out from Hawaii, however, he had a more severe attack of erysipelas that refused to clear up. In the first week of September he became very unwell and eventually took to his bed. He was tended by the two surgeons on board; like all doctors at the time they were more decorative than useful, but they did at least set up a sick-chamber for him on the main deck, where he would have more light and air than in his cabin. Without antibiotics, however, they could do nothing to check the septicemia that was now spreading through his body. On 11 September he became delirious and spoke only in German, even to Maclear, who had become his closest friend on board. After suffering horribly, with nobody able to offer relief, he died in the afternoon of 13 September, two days after his twenty-eighth birthday. He was buried the following day after the custom of the sea, lowered into 2,000 fathoms while his colleagues lined the rail.

Buchanan wrote to his mother to explain her son's sickness and death. One can only imagine her uncomprehending grief—she would only just have received his last letter, dated 14 August, which gave no hint of impending tragedy. She wrote back to ask for more information and Buchanan did his best to reassure her, quite properly concealing how her son had actually died. No doubt Buchanan conducted this correspondence because he spoke German (much of the scientific literature at the time was in German), but Wyville Thomson's silence is a little strange. He was the scientific director of the expedition, after all, and had himself recruited Willemoes-Suhm as a naturalist. It was plainly his responsibility to inform Willemoes-Suhm's mother of her son's strange and unexpected death, yet he delegated this disagreeable task to Buchanan. He did, however, send a short obituary to the journal *Nature* from Tahiti. He and the other naturalists clubbed together for a memorial plaque that was meant to be put up in Gluckstadt, the town in Schleswig-Holstein where Willemoes-Suhm was born. In fact it seems to have been erected in a graveyard in Itzehoe, 10 miles to the north, but afterward it disappeared for many years and was only recently rediscovered at the family burial ground in Segeberg, about 30 miles to the east, near Lubeck; perhaps it had been taken there by his mother. There is talk of moving it again, perhaps to a museum; but for the present the memorial to Rudolf von

Willemoes-Suhm, set up by Wyville Thomson, Murray, Buchanan, Moseley and Wild, can be seen in a graveyard near St. Marien Church in Bad Segeberg.

Station 276: 16 September 1875

13°28′ S, 149°30′ W; Penrhyn Basin, 2,350 fathoms

The ship continued her way across the abyssal plain of the central Pacific, the largest habitat on Earth. Her route cut across the western ends of four huge scars in the seabed, the Clarion, Clipperton, Galapagos, and Marquesa fracture zones, which run parallel to one another for thousands of miles. They are quite narrow, though, and no sounding detected these channels of jumbled rock. Three days after the death of Willemoes-Suhm the trawl was

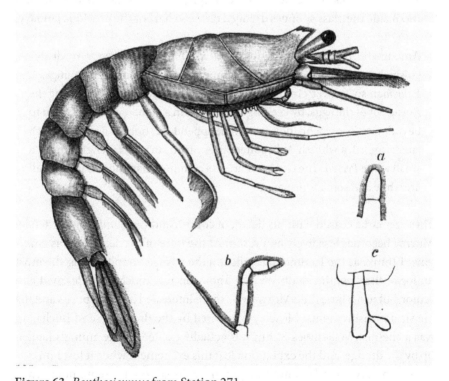

Figure 63 *Benthesicymus* from Station 271

Source: C. Spence Bate (1888), Report on Crustacea Macrura. Zoology, Part 52, Plate 58. Bound in Reports, vol. 24. Also in Narrative, vol. 1, Part 2 Figure 185.

back in action and brought up a large quantity of red clay and about half a ton of manganese nodules. These nodules had long since ceased to be a curiosity and were by now a regular item in the trawl, one whose extent was becoming apparent. The radiolarians had disappeared from the deposit, but there were a few animals: a new ophiuroid, an ostracod, and the same strange big-headed fish they had caught 10 days ago. There was also a big bright red prawn, later given the uncomfortable name *Benthesicymus*, which they had seen a dozen times before, at all depths from 300 fathoms down. Decapod crustaceans such as crabs, shrimps, and prawns are very common in shallow water, but they become less abundant with depth and were only rarely found below 2,000 fathoms, so *Benthesicymus* was an exception. It was also a little puzzling; besides being as red as if it had been freshly boiled, it was six inches long. What did it find to eat down there?

One of Willemoes-Suhm's tasks had been to keep the Station-book, recording what the dredge or trawl brought up and often making enthusiastic technical remarks. Here he is, for example, commenting on the animals found inside the glass sponges dredged off Zebu harbor the previous January:

> Among the parasites inhabiting *Euplectella* the following were distinguished; 1, *Aega spongiphila*, the commonest inhabitant of the Sponge; 2, *Palaemon* sp. (?), very transparent and delicate (I succeeded in getting the Zoeae out of the eggs, by keeping the mother in a globe, and found them to be ordinary Zoeae with, however, some appendages only, as a rule, seen at a later stage); 3, a whitish Aphroditacean Annelid, an inch in length; and 4, a small white *Pecten*. The two latter are less common than the two former; all are white and some transparent.

These remarks ceased with his death, of course, and the *Summary* written by Murray becomes less lively as a result. At the time of his death he was emerging, I think, as the leading naturalist on the voyage, despite being the most junior of the scientific staff: Wyville Thomson was much more learned and senior, but just a little dull; Murray was more interested in the deposits and the physical measurements; Moseley was bored by the dredging; and Buchanan was a chemist. Willemoes-Suhm was actually excited by the animals hauled up by the dredge, and the expedition lost this excitement when it lost him.

The Society Islands came in sight at daybreak on 18 September. They were not part of the original itinerary, but were an appropriate landfall nevertheless, as the society after which they were named was the Royal Society, which had incubated the expedition. Everyone was charmed by Tahiti, as everyone is, and they stayed there for the next two weeks, exploring the hills and charting the reefs. Balfour and Murray captured the monsters—"eared animals"—of the bottomless Lake Waihirra, high in the hills high two days out of Papiete; the lake turned out to be at most 60 feet deep and the monsters were gigantic eels, as long as a man. Moseley found a trout stream where the fish, if not actually trout, would at least rise to a dry fly, and so was able to exercise one of the fly rods he had brought with him. A marine named Leary, who generally accompanied the naturalists on shore, went in for a different kind of fishing: he tied a lump of meat to the end of the line and treeled it slowly past the burrow of a land crab. The crabs are normally very wary and difficult to catch, but they found it difficult to resist the meat. Alas, when they ventured out to snatch it and then rush back to safety they found the mouth of their burrow had been blocked by the foot of Leary.

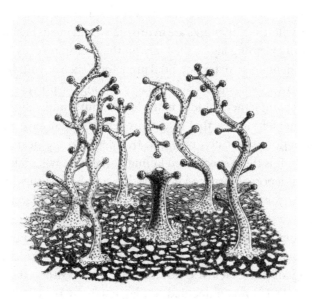

Figure 64 *Millepora* (detail: central gastrozoid flanked by dactylozoids) from Station 276.

Source: H. N. Moseley (1881), Report on Certain Hydrozoan, Alcyonarian and Madreporian Corals. Zoology, Part 7, Plate 14. Bound in Reports, vol. 2.

Fire coral

Perhaps I have been too hard on Moseley, who came into his own when the water was shallow enough for him to mess about in a small boat. Dredging was very difficult on the reef, but he persevered long enough to build up a collection of corals. Corals were the animals that interested him most, and after the voyage he was responsible for describing the specimens that had been collected, many of them by him, dredging in the ship's boat or the steam pinnace off the shores of the islands they visited. He collected 40 or so species of coral at Tahiti. "Coral" is a colloquial term, however, that describes a growth form rather than a particular kind of animal. The corals that build reefs are mostly hexacorals, named for their six-rayed internal structure, that sit in stony cups like tiny sea anemones and cultivate their symbiotic algae. A few reef corals are alcyonarians, soft corals like the pennatulids and gorgonians trawled from deep water, which have acquired the knack of secreting a stony coat to make a hard soft coral, so to speak. Moseley was particularly taken by the millepores, the fire corals that you are shown when you first visit a reef and told never to touch with your bare skin. These are quite different: hexacorals and alcyonarians are both anthozoans, whereas millepores are hydrozoans, and mean business. Rather than peacefully farming algae, they rely on their stinging cells to catch live prey and incidentally to deter potential predators (such as your hand). What fascinated Moseley was the division of labor among the polyps. Some have a conventional complement of four tentacles, a mouth, and a digestive cavity. There are others, however, that bear more tentacles but have no mouth at all; their function is to catch prey and transfer it to the nearest neighbor with a mouth. It is these specialized killing polyps that will make you regret touching a millepore. More zoologically, they are the first step in the evolutionary series from a simple colony toward a compound differentiated individual such as a siphonophore.

The day before leaving Tahiti, Wyville Thomson decided to invite a party of local dignitaries on board for a short cruise to demonstrate the working of the scientific equipment. The visitors included royalty, so Wyville Thomson decided to play it safe by trawling outside the harbor in 680 fathoms, from which he could confidently expect a rich haul. The sea was whipped up by a strong trade wind, however, and the day was rough and squally. Worse, the trawl came up empty, except (according to Murray) for a small shrimp clinging to the netting. Perhaps it hoped to be introduced. The following day, 3 October, *Challenger* reluctantly left Tahiti for the 5,000-mile transit of the South Pacific to Valparaiso.

Challenger steamed out of Papiete harbor, to avoid the changeable land winds, before settling down to sail south and then east for the coast of South America. The seabed was undulating, 2,000 to 2,500 fathoms deep, with red clay changing to ooze above about 2,250 fathoms. They soon lost the trade winds and were frustrated by light airs that barely stirred the ship, which had to resort to steam. The general grumpiness of sailing slowly across an illimitable sea returned. Matkin was out of sorts; he had received no mail in Japan, nor in Hawaii, nor in Tahiti, and felt forsaken. Campbell has nothing to say about this leg of the voyage except that it was long and tiresome. Moseley passes over it in silence. It was left to Murray to record one of the strangest and least expected discoveries of the expedition, a mass grave of great fish and whales.

Station 281: 6 October 1875

22°21′ S, 150°17′ W; Austral Seamounts near Rurutu Island, 2,385 fathoms

Three days out of Papiete the trawl was recovered successfully with a lot of red clay and a few animals, mostly sponges. There was also a deep bathysaur, perhaps the deepest predator found on the voyage, and one of the odd legged holothurians peculiar to the deep sea and first found by the *Challenger*. Like other sea pigs it ambles across the surface of the sediment, eating what it finds like an animated conveyor belt. This particular animal, *Oneirophanta*, is about as big as your hand and can be extraordinarily abundant, reaching densities of 30 or so individuals per hectare (squirrels in Montreal come to mind) and a biomass equal to all the big animals combined. Oddly enough, there seem to be no males in some populations, where about half the adults are female and the other half are neuters, with no evident gender, which is difficult to explain.

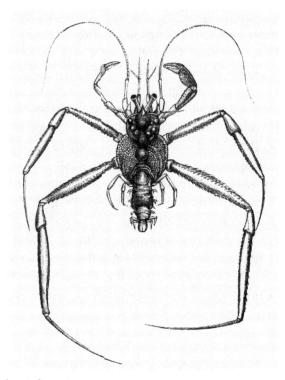

Figure 65 *Tylaspis* from Station 281

Source: J. R. Henderson (1888), Report on Anomura. Zoology, Part 69, Plate 8. Bound in Reports, vol. 27. Also in Narrative, vol. 1, Part 2 Figure 329.

The homeless crab

A week later and almost 500 miles to the southeast, neither the depth nor the deposit had changed much. The trawl brought up a few animals, including *Tylaspis*, the deepest crab they found. It is a hermit crab that has lost its vocation. All other hermit crabs have a soft, uncalcified abdomen that they protect by stuffing it into an empty snail shell. This accounts for the snails (as they appear to be) that you see ambling along quite fast in a rock pool at low tide. *Tylaspis* alone does not use gastropod shells and has instead begun to rearmor its abdomen, just like a real crab. There may be a simple reason for this: only two gastropod shells were trawled from water as deep as this during the entire expedition, so the ancestors of *Tylaspis*, faced with being homeless, adopted the only other way of protecting their rear end.

Station 286: 16 October 1875

33°29′ S, 133°22′ W; southwest Pacific Basin, 2,335 fathoms

Challenger was now sailing roughly southeast over a rather uniform abyssal plain about 2,500 fathoms deep in the middle of the South Pacific. This is one of most remote places on Earth, thousands of miles from the nearest mainland. The trawl brought up a few animals, although not very many, usually a dozen or so. What it did bring up in great quantity were the remains of the biggest animals that live in the ocean. At the station where they found *Oneirophanta* they also found over 100 sharks' teeth and six ear bones of whales. (The ear bones, about the size and shape of a clenched fist, are made of the hardest and densest bone and last longer than any other part of the skeleton.) The trawl that brought up *Tylaspis* also brought up more than 1,500 teeth and about 50 ear bones. At the next station, there were over 300 teeth and about 90 ear bones. A week later only a couple of teeth and ear bones were found, and after that they turned up occasionally, but never again in such abundance. The trawl had swept a corridor 500 miles long in the middle of the Pacific Ocean that was the burial ground of great fish and whales.

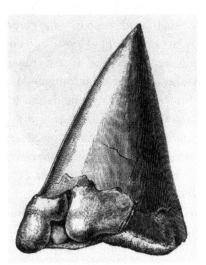

Figure 66 Shark's tooth from Station 286".

Source: W. Turner (1880), Report on the bones of Cetacea. Zoology, Part 4, Plate 2. Bound in Reports, vol. 1. Narrative, vol. 1, Part 2 Figure 294.

Whalefall.

Phlebas the Phoenician, a fortnight dead,
Forgot the cry of gulls, and the deep sea swell
And the profit and loss.
                    A current under sea
Picked his bones in whispers.
                              —T. S. Eliot, *The Waste Land*

But the current would have been slight; and others would have picked his bones before. A dead whale imports a huge mass of tissue to the deep sea from the surface, and this traffic has been going on since the ancestors of modern whales set out for the open sea 40 million years ago. This has given plenty of time for a distinctive community of animals to evolve: the necrophages, or corpse-eaters, who wait patiently in the dark for the first signs of a whalefall. Those who arrive first get the finder's share, and the faster they eat, and the more they eat, the greater their share will be. Necrophages are therefore large active animals with highly developed senses and enormous stomachs. The first to arrive in shallow water are usually hagfish and sleeper sharks, which can tear large pieces from the corpse. In the deep sea the first to arrive at the feast are often amphipods, the universal necrophagous stomachs of the deep sea, burying their heads deep into the corpse and gorging themselves; a good meal may last an amphipod a year. I do not know what happens to Phoenicians, but pig carcasses, used as a surrogate, are crawling with amphipods within a few hours. They are accompanied by macrourid fish, the grenadiers, and large prawns, which may be partial to whale flesh but also attack the amphipods (so that is what *Benthesicymus* was eating). The giant amphipods of still greater depths may have evolved large size primarily as a protection against these predators. The regular arrival of dead whales on the sea floor stimulates the assembly and evolution of a complex community of large animals specialized to consume them.

The necrophages are capable of stripping 100 pounds of flesh from the carcass every day, and even a large whale will be skeletonized within a few weeks or months. When they are finished, an extensive patch of polluted sediment remains, which is colonized by waving fields of polychaete worms that harvest the bacteria. Crabs and ophiuroids will join the fish and prawns to eat the eaters. And even when they are done, the slow decay of bone lipids will provide a food source for mussels, limpets, and worms

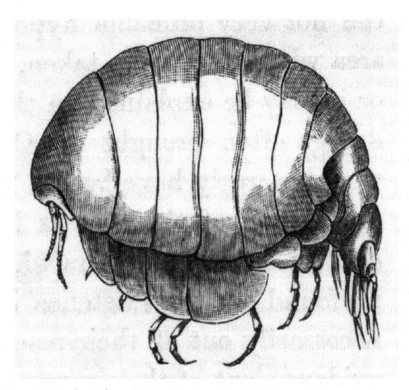

Figure 67 *Andania* from Station 67
Source: Narrative, vol. 1, Part 2 Figure 207.

able to use sulfur compounds in solution. One of the gutless worms, the pogonophorans, that the *Challenger* failed to collect uses its symbiotic bacteria to suck the juices from whale skeletons. Eventually, a decade or so later, the very last edible scraps have been consumed. The whalefall, a hill of flesh on the sea floor, has been consumed completely by animals that must now move on to find their next meal, leaving only the ear bones behind as witness. This turns our notion of deep-sea ecology on its head. We had imagined a thin uniform rain of surface plankton delivering starvation rations to a patient community of sessile filter feeders. The extensive graveyards of whales trawled by the *Challenger* suggest a completely different picture: the occasional arrival of a feast that will initially be consumed by large, active necrophages and subsequently persist for years as a patch of enriched sediment supporting a complex community of animals, all of which eventually depend on whalefall.

Something like this must have happened repeatedly to produce the tracts of whale bones on the sea floor dredged up by the *Challenger*, which cast some doubt on the idea that the deep sea receives its subsidy from the surface largely as a thin rain of small particles. At the surface, the combined mass of organisms within some given range of size is a constant, so the biomass of bacteria is (roughly) equal to the biomass of whales. This rule does not hold in the deep sea, however, because more massive objects, such as dead whales, sink faster than small objects, such as bacteria (Galileo's demonstration that large and small spheres dropped from the leaning tower of Pisa fell to the ground at the same time would not have worked had he been a marine animal.) It might seem reasonable to think that objects that sink faster will pile up at the bottom of the sea, but this is not so. Imagine a race-track a mile long with two lanes. One lane is for runners, who run at 10 mph, and the other for walkers, who walk at 3 mph. Every minute, one runner and one walker leave the starting line, set off down the track, and are immediately replaced by another runner and walker (that is, there is an equal supply of runners and walkers). After six minutes, the first runner crosses the finish line, while the first walker has covered less than a third of the distance. After one more minute, the next runner (having started out a minute behind the first) finishes, with the first walker still plodding along far behind. It is only after 20 minutes that the first walker finally arrives at the finish; but the next walker is only a minute behind, and from that point on, one runner and one walker cross the finish line every minute. Once the system has settled down, the rate of arrival at the finish is the same, regardless of speed along the way.

What matters is not sinking speed itself, but sinking time: a large object, sinking rapidly, will have less time to decompose and will deliver proportionately more nutritive biomass to the sea floor than the same mass divided up into small, slow-sinking particles. (Large objects may decay more slowly anyway because they have relatively less surface area, but animals with guts decay from the inside out as well as from the outside in, so the calculation is complicated.) The water column acts like a sieve that selectively removes smaller particles. The result is that the flat distribution of biomass at the surface becomes tilted toward larger size categories on the seabed. Some simple calculations suggest that in the deep sea it may become very strongly tilted, so that most of the usable organic matter arrives as large dead animals: a whale arrives almost intact, whereas only the husks of planktonic bacteria, algae, and forams survive their long descent.

One final twist is that industrial whaling during the nineteenth and twentieth centuries reduced large whales to about 10 percent of their original
abundance, and no doubt the animals that fed on their carcasses were proportionately depleted. (The widespread finning of sharks will have a similar
result in a few years' time.) Perhaps this was counterbalanced to some extent
by warfare; 40 years after the *Challenger* returned to port, the Royal Navy
lost about 40,000 sailors in World War I, but it is difficult, as well as rather
ghoulish, to make any detailed calculation. It seems very likely, though, that
human activities have strongly affected the deep seabed that until recently no
human eyes had ever seen.

Station 289: 23 October 1875

39°41' S, 131°23' W; southwest Pacific Basin, 2,550 fathoms

A week later, making little more than 50 miles a day in light southerly and easterly winds, the ship was leaving the whales' graveyard behind and the trawl
was bringing up new animals. One of the most remarkable was *Nebaliopsis*,
belonging to a small group, the phyllocarids, that stands apart from all the
other crustaceans—amphipods, isopods, shrimps, crabs, and so forth. It is
an inch or two long and swims above the deep sea floor. It looks something
like a shrimp in a large cloak, the carapace that encloses most of its body,

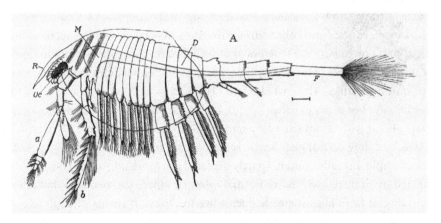

Figure 68 *Nebalia* from Station 289
Source: Narrative, vol. 1, Part 1 Figure 63A.

leaving only the end of its tail sticking out. The fossils of animals that look like phyllocarids have been found in Cambrian rocks, soon after the great radiation of animals, so *Nebaliopsis* has been added to the list of living fossils found in the deep sea. Most of its relatives, however, are scavengers living in rather shallow, muddy water, which doesn't seem to fit the theory at all.

That sinking feeling

There were also several new glass sponges in the trawl, as there usually were from deep water. These are, as I've described, animals that are highly specialized to filter bacteria from seawater. But where do the bacteria come from? The conventional explanation, one that I've taken for granted till now, is that they sink slowly from the productive zone at the surface of the sea. The flaw in this scheme is that they must sink very, very slowly indeed. If you have ever taken basic physics courses you may have a dim recollection of learning about Stokes' Law, which describes how fast small particles sink through a fluid. A bacterium is a very small particle, usually about one micron, one-thousandth of a millimeter, in diameter. According to Stokes, an inert bacterial cell will take nearly two minutes to sink its own body length in seawater, and would therefore take 4,000 years to sink to 2,000 fathoms, except that it would have completely decomposed long before that. We can try to rescue the theory by supposing that bacterial cells do not sink separately, but in groups, or attached to flocs of organic matter, or in fecal pellets, all of which will sink faster. This notion has often been suggested, and midwater particle traps show that there is something in it. The mucus feeding webs shed by pteropods and other surface dwellers may carry viable bacteria and algal cells downward at 50 meters or more a day, reaching the bottom in deep water in a few weeks and perhaps scavenging smaller particles from the water column as it sinks. The cast skins of crustaceans sink quite fast, at least to begin with, but seem to break up quickly and are seldom found in midwater traps. Fecal pellets can sink very rapidly, although estimates are all over the place, and they do not necessarily remain intact for long. A fresh salp pellet, for example, initially contains partly digested phytoplankton, but is soon colonized by bacteria, which are in turn eaten by ciliate protozoa, so that after four days it is an almost sterile floc of marine snow. It seems difficult to get past the conclusion that only the largest objects will reach the deep seabed in good shape.

If there is anything in this, we would not expect there to be many bacteria or other microbes, such as algae or fungi, at great depths. (The anecdote that is often cited to confirm this is the immortal bologna sandwich that was left in a lunchbox in the *Alvin*, a submersible that sank in 750 fathoms in 1968. When the vessel was recovered 18 months later the sandwich had scarcely decayed, although I don't think that anyone actually ate it. It had been at refrigerator temperature all this time, of course.) Yet animals that are specialized to eat bacteria certainly live at great depths, not only glass sponges but also bryozoans, brachiopods, and others. This seems paradoxical. Let us try to imagine in more detail what happens during a whalefall.

Plume and splash

A dying whale, old and in poor condition, will sink because it has more bone than blubber. As it sinks, struggling feebly, it will accelerate until its buoyancy balances gravitation and it reaches terminal velocity. Stokes' Law doesn't help because it applies only to very small objects, but from more general principles, and making several more or less plausible assumptions, I calculate that a spherical whale of moderate size will have a terminal velocity of about three meters per second, about jogging speed. Since whales are streamlined rather than spherical it will in fact fall faster than this, perhaps as much as four or five times faster, at least if it falls head first. It will descend quite soon to a depth where pressure forces the remaining air from its lungs and it dies. Further down, beyond the depth to which whales can dive, it will be squeezed like a toothpaste tube and will continue to sink, dead, trailing a thin plume of tissue and gut contents. When it reaches the bottom of the sea, only an hour or so later, it will plow into the sediment with the momentum of a bus traveling in a school zone until brought to a stop. A cloud of sediment, mixed with the resident animals and fragments of whale tissue, will billow upward and outward before very slowly settling on an area of sea floor about the size of a hockey rink.

The plume left by a sinking whale and the cloud raised when it splashes down will inoculate the deep water with bacteria—intact, viable bacteria rather than the husks and fragments that sink from the surface. Using a chain of rather dubious assumptions and approximations, I think that a whale might immediately deliver about 10^{18} bacteria, enough to pollute about 10 million cubic meters, or 100 million cubic feet, with enough bacteria to interest a sponge. The corpse will then start to decompose as its resident bacteria begin to proliferate; after all, a beached whale soon becomes a public health problem, even in winter. Not all the bacteria will be able to grow at high pressure and low temperature, but some will, expanding to take the place of those that cannot. No bologna sandwich will survive. A good deal of the flesh of the whale will be eaten by the necrophages before it can decompose, but the necrophages themselves will digest their food and excrete the remains, mixed with bacteria. When the bulk of the corpse has gone, the bacteria will remain in the polluted sediment, where they are consumed by xenos and komokis, which in turn are eaten by deposit feeders such as holothurians, echinoids, and worms. As these push through the sediment or trundle over its surface, they will inevitably displace puffs of detritus and bacteria; the minute size that prevents them from sinking fast from the surface now serves to keep them in suspension above the sea floor, where they will long continue to provide a source of sustenance for filter feeders.

One way of summarizing all of this is to say that the ecological wiring at the surface and on deep sea floor run in opposite directions. The base of the food chain at the surface of the sea are innumerable unicellular algae that are eaten by small crustaceans that are eaten in turn by jellyfish and arrow worms, and so forth up to adult whales and sharks, which are not themselves eaten by anything. On the deep sea floor it is these whales and sharks that form the base of the food web, with a whole community relying directly or indirectly on their flesh and the bacteria they bring with them. This, at least, is the conclusion of a long (and in places fragile) chain of reasoning that begins with the bones that *Challenger* scraped from the seabed in the middle of the Pacific Ocean.

Station 295: 5 November 1875

38°07' S, 94°04' W; Chile Rise, 1,500 fathoms

The ship had made good progress under sail until 17 October, but was then headed by the wind and forced south. The westerlies were picked up a few days later and allowed Thomson to head more or less directly east along the 40 degrees south line of latitude. The sea became shallower toward the East Pacific Rise, and the bottom changed from red clay to ooze. The weather was fine, although becoming much cooler, and too calm. The great enemy

Figure 69 Minute black metallic spheroid from Station 295
Source: Narrative, vol. 1, Part 2 Figure 295.

of sailing ships is not the storm, which can be ridden out given enough sea room, but the calm, when nothing can be done but loaf along, at best, until the wind returns. *Challenger* was becalmed for about a week, from 3 November, but of course she could do something about it and steamed for 200 miles. A breeze sprang up to revive their hopes on the eighth and the sails were set, but it died away and the engine was started up again for the rest of the way to Juan Fernandez.

Extraterrestrials

Six attempts were made to trawl during this weary time. The trawl was lost once, with 1,600 fathoms of line, and once came up foul. Three trawls produced only a one or two animals, and only one brought up a reasonable number of specimens. They were traveling across a desert of red clay. While the other naturalists languished, Murray consoled himself by taking a closer look at the sediment. In the middle of the Pacific, far from land and any human influence, only red clay accumulates below the carbonate compensation depth where the shells of forams and pteropods dissolve. It accumulates very slowly indeed, at a rate of only two or three millimeters in a thousand years. During my lifetime or yours, the clay on the bed of the abyss will build up by no more than the thickness of a sheet of paper, so the sediment that has accumulated since the last retreat of the glaciers is no thicker than the book you are reading. Murray was surprised to find that the clay often contained dozens of tiny spheres, each a fraction of a millimeter across. They could be collected with a magnet and were clearly made of magnetic iron, often with some nickel, copper, and cobalt. What were they? They resembled the particles from welding shops, furnaces, and locomotives, but the Industrial Revolution had only just got into its stride and the nearest factory was thousands of miles away. In any case, some of the mysterious spheres were well below the surface of the sediment. Volcanic debris was widespread—the sea floor hereabouts was strewn with pumice—but this usually consisted of irregular fragments of glass or stone. Having ruled out the obvious terrestrial sources, Murray guessed that they were extraterrestrial. The cosmic dust sucked in by the Earth's gravitational field, or the particles ablated from meteorites, would be melted during their descent and solidify again as spheres. Their composition resembled that of an iron meteorite. It seemed reasonable enough, but it was an audacious guess. Within living memory, Thomas Jefferson, hearing of a meteorite impact in Connecticut, had famously remarked that he would sooner believe that two Yankee professors would lie than believe that stones fall from heaven. That sounds reasonable too, but Jefferson was wrong and Murray was right. The thin film of red clay, accumulating with exquisite slowness deep in the ocean and far from land, provides a canvas on which these tiny spheres are easily discerned—provided that you look. Murray looked, and discovered the witnesses of the continual bombardment of the earth by miniature meteorites whose existence nobody had previously suspected.

The waters above

They found very few animals in this desert of red clay, larger by far than the Sahara. There seemed to be even fewer in midwater, which was searched by the deep tow-nets. These were the same nets that had been used since the start of the voyage to sample the plankton near the surface of the sea and were often crammed with the algae, larvae, and small animals that swarm in the sunlit waters within a few hundred feet below the surface. Willemoes-Suhm had speculated that pelagic animals might live at much greater depths, and the previous April, on the way to Japan, the naturalists had begun to experiment with tow-nets fished much deeper. Between Japan and Hawaii this became a regular practice. It was not very fruitful at first: the trawl fouled, or nets attached to the trawl filled with clay. Eventually, however, tow-nets were spaced out along the trawl line and began to produce results. They caught a lot of radiolarians, including very large colonial species, and a few animals, including deepwater hatchetfish and bright red shrimps. For the most part, however, nets that were towed below 500 fathoms or so did not catch very much. Wyville Thomson concluded that

> there is every reason to believe that the fauna of deep water is confined principally to two belts, one at and near the surface, and the other on or near the bottom; leaving an intermediate zone in which the larger animal forms, vertebrate and invertebrate, are nearly or entirely absent.

Forbes' azoic zone had returned, displaced upward. What Wyville Thomson is asserting is that animals may live at any depth on the bottom of the sea, or just above the bottom, but not in the sea itself between the surface zone and the seabed. The surface zone extends into the twilight 500 fathoms below the surface, where there is an abundance of large animals—there has to be, since sperm whales and elephant seals dive this deep to feed. In deeper water, according to Wyville Thomson, very few animals can be caught. He has been strongly criticized for this conclusion, but in one important sense he was not far wrong. Imagine that you are in a large public room, the Central Hall of the Natural History Museum in London. It will hold over 1,000 people (it often does, at weekends) and is dominated by a huge Blue Whale, slung from the roof 60 feet above. Now imagine that the whale is no longer there, all the people have gone, and all the other objects and displays, even the statue of Richard Owen, the antagonist of Darwin, have been removed to storage. In place of the whale there are two shrimps and a small fish the length of your thumb. This will give you some idea of the emptiness of the largest room on Earth, the world ocean between the twilight zone above, and the sea floor far beneath. Animals are very scarce. With the primitive equipment that the *Challenger* could deploy Wyville Thomson's conclusion seemed justified, and even modern towed nets, opening and closing at specified intervals and depths, largely bear him out. Except for the jellies.

Towed nets are not kind to gelatinous animals, whose delicate bodies are usually ripped apart or badly mangled when they are rubbed against the mesh. Midwater tows recover robust organisms such as fish and shrimps while reducing gelatinous animals to unidentifiable lumps and fragments.

It was only when remotely operated cameras began to be deployed in the deep sea, a century after the *Challenger* sailed, that the abundance and diversity of the animals that live there was revealed. Many of them are gelatinous sandwich-animals with an external layer of cells separated from an internal layer by a soggy middle layer of connective tissue. Siphonophores, with their long washing-lines of zooids, are no longer mere curiosities but dominant predators. Huge medusae drift past in the dark. Ctenophores with their banks of miniature oars row past, searching for prey. All of these have gelatinous bodies, and you might wonder how such primitive creatures have managed to survive to the present day. Think again. That boring layer of connective tissue that glues their bodies together is in fact a composite material with remarkable properties. Like other composite materials, such as fiberglass, is consists of fibers (of collagen) embedded in a matrix (of polysaccharide), and it is highly extensible and elastic. Any shrimp with haughty opinions of its superior state is likely to change its mind when it is captured in a forest of tentacles and towed toward the waiting mouth.

It is possible to compile a long list of the advantages of having a gelatinous body. It is cheap to make; it can grow quickly; it offers little nourishment to potential predators; it is easily made transparent; and so forth. But perhaps it is not necessary to identify particular advantages. Perhaps it is the default option for animals. In special circumstances, where waves or tides or currents or gravity make it necessary, a robust body will evolve. The animals with which we are most familiar, in the forest or in the river or at the surface of the sea, need armor and skeletons to cope with the physical stress of their daily lives. In the calm of the deep sea these are extravagances, and many of the animals that live there have adopted the most natural configuration, the gelatinous body.

Station 298: 17 November 1875

34°07′ S, 73°56′ W; western flank of Peru-Chile Trench, 2,225 fathoms

The hauls on this section of the voyage had been scanty but produced a few animals. One was a snipe eel, one of the midwater fish that was probably caught as the trawl was being recovered. As its name suggests, it looks like an eel that has been unexpectedly provided with extraordinarily prolonged jaws. Another was a stony coral that lives below the depth at which a calcium carbonate covering can be secreted, so it has given up and lives naked, looking for all the world like a sea anemone. They brought up a sea urchin that decorates its test as though it were a spider crab, plastering it with xenos and whatever else it can find on the sea floor. On the whole, though, the slow crawl over the abyssal plain was unrewarding, the weather was wet and chilly, and they were relieved to sight Juan Fernandez, the Robinson Crusoe island, on the morning of 13 November. The ship anchored in Cumberland Bay, in rather uncomfortably deep water overlooked by towering cliffs, and stayed there for a couple of days. They visited Alexander Selkirk's lookout, of course, where he had spent four lonely years scanning the horizon for the blink of a sail, and dined like him on feral goat, which Campbell found excellent. They did not stay long, though, and headed out into rough seas in the evening of 15 November.

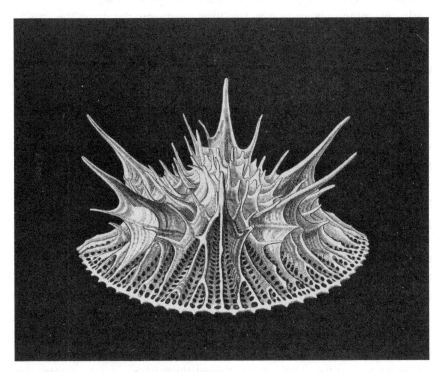

Figure 70 *Leptopenus* from Station 299
Source: Narrative, vol. 1, Part 2 Figure 289.

The usual suspects

Two days later found them on the lip of the long trench that runs parallel to the coast of South America. Here, at last, the trawl brought up a respectable number of animals. Most of the species were new to science, but were very similar to species they had collected before elsewhere in the Pacific, or in the Atlantic or the Southern Ocean. There was Willemoes-Suhm's blind lobster, described by him as *Deidamia* but afterward fittingly renamed as *Willemoesia*. Moseley described the beautiful deepwater coral *Leptopenus*, looking like a lacework cup sitting upside down on the seabed, with its long delicate spines extending upward. There was also a clutch of deep-sea holothurians, including the ghostlike *Peniagone* and the peculiar finned *Psychropotes*. Other deep-sea echinoderms included the "snorkel starfish" *Porcellanaster* and the slipper-shaped sea urchin *Pourtalesia*. They had seen the ophiuroid *Ophiacantha* and the pycnogonid *Colossendeis* several times before. What they had discovered was not merely a lot of previously undescribed animals, but rather an entire distinctive community of animals that lived together in the abyss. The expedition has often been criticized for its feeble, inadequate scratching of the seabed, but no enterprise before, except perhaps for the invention of the microscope, has succeeded in revealing a whole new world of organisms whose existence had previously been unpredicted, or at best had only been guessed at.

The ship made landfall at Topocalma Point and steamed north to anchor in Valparaiso on 19 November. Campbell and Balfour had been promoted and left the ship to return home, or, as Campbell put it, HOME! Rather than tamely taking passage in another ship, they set out on horseback despite expert advice that scaling the Andes and riding across the pampas on the far side, with no Spanish beyond ordering a beer, was a foolish enterprise. They crossed the continent anyway and sailed from Montevideo early the following February. Half a dozen seamen left the ship at the same time, melting away into the town, which was then strongly influenced by British commerce and quite prepared to welcome them. Matkin had received letters from home at last and was in better spirits, although he complained bitterly that he had to do the work of the ship's steward, who had fallen ill in Tahiti, without extra pay. The Paymaster, Richards, told him that it was a great honor, but he would clearly rather have had the money. Everyone else had three weeks in which to refit the ship and enjoy themselves in Valparaiso and Santiago before facing the last obstacle, the intricate passage that led to the Strait of Magellan and through there to the open Atlantic.

17

Twelfth Leg

The Patagonian Fjords, December 1875–February 1876

Full Fathom 5000. Graham Bell, Oxford University Press. © Graham Bell 2022.
DOI: 10.1093/oso/9780197541579.003.0018

There were two options, then as now. The first was to go outside the islands and risk the stormy Pacific, trusting that the westerlies would blast the ship round Cape Horn. The second was to pick their way slowly through the Inner Passage and enter the Atlantic by way of the Strait of Magellan. The first was faster, the second safer. Thomson chose the safer route, as the Admiralty had specified. He had to make a dogleg first, however, forced by strong southerlies to beat far out from the coast, back toward Juan Fernandez. A trawl at nearly the same depth as the previous station brought up a very similar collection of animals. The fourth Christmas of the voyage was celebrated decorously with a concert and recitations. The Captain favored the crew with a violin solo accompanied by the piano that he had brought on board in Hong Kong. The crew, however, had managed to smuggle a great deal of drink on board in Valparaiso, and before long the party became much less decorous, indeed positively riotous. One seaman had his jaw broken in three places, according to Matkin. He explained the indiscipline as the natural consequence of the plague of desertions: when a passing naval ship was implored to send a few men to fill the vacancies, it would often take the opportunity to offload its most troublesome characters. Other accounts pass over this Christmas Day in silence; and there was no dredging or sounding on Boxing Day.

A week later they caught the westerlies and returned to the coast at Tres Montes, at the mouth of the Gulf of Penas, which they sighted in the morning of 31 December, and sailed a few miles up the coast before anchoring on the sandy beach at Port Otway (Puerto Almirante Barroso). The scientists and officers sat up late that night to drink in the New Year (decorously, I presume) and hear 16 bells rung, eight to mark the passing of the old year and eight to herald the new. The ship sailed early the following morning, New Year's Day, across the bay to the entrance of the Messier Channel, which would lead them south toward the Magellan Strait. This part of the voyage was in complete contrast to the open Pacific. The inner channel lies between the mainland and the outlying screen of islands, hemmed in by steep cliffs rising to a thousand feet or so and clothed down to the water's edge with the vast impenetrable forest of Patagonia. It is nowhere more than two or three miles wide and often considerably less, and the cliffs kill the wind, so that a large sailing ship can find it difficult to wriggle through. The scenery might be described as romantic and primeval, or as a dismal wilderness glimpsed through driving rain, according to the point of view; but it did give frequent opportunities for exploring the harbors and islands. Moseley would leave the ship in the afternoon to crawl through the undergrowth in search of birds to shoot, but often he could make very little progress through the dense tangle of stems and branches.

Picking their way through the winding channel was also a good deal more adventuresome than sailing the open ocean. Two days later they stopped in Gray Harbour, where a large lake at the head of the bay gave a welcome fishing-and-hunting break for the officers, while the men went on shore to wash their clothes. Naturally, they lit fires to dry them, and these set the dry grass around on fire, which spread to the forest, which despite its dampness was soon on blaze. Nobody worried too much about this—Spry found it "a grand sight"—but they did worry when Wild, the artist to the expedition, failed to return to the ship at 6:00 p.m. with the boats. Darkness was falling, the forest was on fire, and Wild was missing. On this tangled unknown shore, with visibility limited by trees and the failing light, Wild was utterly lost and very frightened. By great good luck he found a rocky promontory that gave him some protection from the flames, and it was here that one of the search boats found him two hours later. He had had a narrow escape.

The following day, in wet, squally weather, they found a derelict German steamer at Port Grappler, abandoned by her captain when a rock knocked a hole in her bottom. Four English sailors from a passing Chilean steamer had taken possession, and no doubt earned a year's wages from the salvage; they emphatically refused any assistance from the *Challenger*. The ship went on to Tom Bay, where the same thing nearly happened to them, when, at anchor, the stern swung close to a rock just awash in the middle of the channel, and disaster was avoided only because steam was already up in the boilers. A week later they nearly drifted on shore at Port Churruca when an anchor broke on the rocky ground, but let go a second anchor in time and steamed out of danger. There were many difficulties and the weather was generally vile, but it was better than crawling slowly across the endless wastes of the Pacific.

Station 308: 5 January 1876

50°08′ S, 74°41′ W; Canal Grappler, 175 fathoms

Animal forests

During this unique section of the voyage, *Challenger* threaded her way past the innumerable islands, bays, and fjords of southern Chile. This is the most extensive fjord system in the world, but even today it is little known, and is still coming up with surprises. This was a far cry from their Pacific cruise. Instead of the deep water of the open sea, the ship was sailing on an inland waterway with a depth of a few hundred fathoms at most, and usually a good deal less, with a rough, rocky bottom that sometimes made dredging hazardous. The great consolation was that instead of the scanty fauna of the red clay the naturalists found animals here in abundance, most of them belonging to species that had never been described before, and some of them very strange indeed.

On 5 January they recovered the trawl from 175 fathoms off Madre de Dios Island and found it bulging with the long, slim stems of gorgonians. It had swept through one of the "animal forests" that clothe the floor and sides of the channel. "Forests" may be a little too grand, since the gorgonians stand only a foot or so tall, so "garden" might be more appropriate, but in any event they provide shelter for a diverse community of other animals. Animals like these gorgonians, and the deepwater hydrozoan corals that we shall meet on the way to the Falklands, have been dubbed as landscape engineers, as though they had

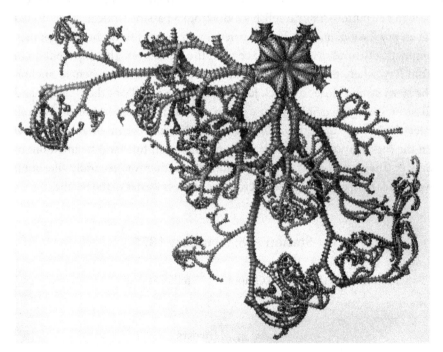

Figure 71 *Astrophyton* from Station 308
Source: T. Lyman (1882), Report on Ophiuroidea. Zoology, Part 14, Plate 47. Bound in Reports, vol. 5.

contracted to supply housing for a list of clients. Nothing could be more absurd, of course. They grow where they can, and a host of other animals take advantage of them. They really do create distinctive landscapes, though, like corals and beavers and black spruce, with an equally distinctive fauna. There was a pretty nudibranch mollusk, for example, later named *Tritonia challengeriana*, which browses on the gorgonians. They missed the extraordinary anemone *Dactylanthus*, which somehow contrives to consume large patches of polyps from a gorgonian before inflating like a toy balloon and drifting off to find another victim. However, they did catch the equally extraordinary gorgon-head, *Gorgonocephalus*, an ophiuroid drawn by Salvador Dalí, with endlessly bifurcating arms, so that a simple five-armed brittle star becomes transformed into a lacy sheet. This is not a decorative structure: the innumerable branchlets are armed with tiny hooks to snag any small animal that wanders into the embrace of the gorgon-head. There was a crinoid, always welcome, a pycnogonid, and a clutch of bryozoans; about 70 species altogether, 40 of them seen for the first time. On land the forest had petered out and the hills were bare rock, but there was plenty of life in the sea, in the animal gardens.

Station 311: 11 January 1876

52°45' S, 73°46' W; Smyth Channel, 245 fathoms

A very unexpected result

Three days later they had sailed 300 miles through the inner channels and were passing south down the Smyth Channel in a thick drizzle. Trawling here produced another monster catch, the net packed with sponges, echinoderms, bivalves, crabs, and tunicates. It also contained something that looked at first glance like a piece of seaweed, the sort of light brown, somewhat translucent straps that makes the rocks slippery at low tide. One glance under a lens showed that it was in fact a colonial animal consisting of hundreds of small zooids spaced out along a richly branched tube. At first it was taken to be a colonial tunicate, but this was soon dismissed. Perhaps it was a hydroid? It was sent to George Allman, the Professor of Natural History at Edinburgh, who didn't think so. More likely it was a polyzoan, and should be described by George Rusk in London. He thought it was a little odd and passed it on to William

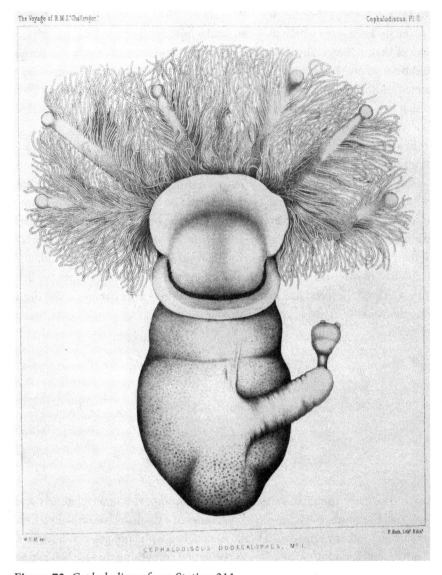

Figure 72 *Cephalodiscus* from Station 311

Source: W. C. M'Intosh (1887), Report on *Cephalodiscus dodecalophus*. Zoology, Part 62, Plate 2.
Bound in Reports, vol. 20.

M'intosh at St Andrew's, who finally agreed to write a report dedicated this single peculiar animal, which he called *Cephalodiscus*. He gave a very detailed description of its anatomy, but in turn passed it on to Sidney Harmer, one of the group of young scientists at Cambridge who around this time were shaking up the venerable field of comparative zoology. At this point the music stopped and Harmer unwrapped the parcel. M'Intosh may have written the report, but Harmer wrote the appendix, and dropped a bombshell.

Over two years ago, if you recall, *Challenger* had dredged up a wormlike animal, *Glandiceps*, from the Atlantic. It turned out to be a hemichordate, distantly but clearly related to chordates, our own group of animals, because it possessed gill slits. To be a little zoological, it had a dorsal nerve cord (in other invertebrates, such as annelids and arthropods, the nerve cord is ventral) and something reminiscent of a notochord, as well as gill slits. Harmer was able to reconstruct in great detail the animal from the Smyth Channel by using the new technique, pioneered in Cambridge, of making serial sections. The specimen is first soaked in molten wax, so that it becomes embedded in a solid block when the wax cools. This block is held fast and cut with a very sharp blade that can be moved up and down by turning a wheel, like the needle of a sewing machine. When the blade is advanced a fraction of a millimeter along the specimen by a screw, it shaves off a very thin section that can then be mounted on a microscope slide. Reducing the specimen to a stack of thin sections allows its anatomy to be reconstructed in great detail, much as we would now use CAT scans to do the same thing on larger animals, such as ourselves. Harmer's serial sections of the Smyth Channel animal showed that it had a dorsal nerve cord, something reminiscent of a notochord, and gill slits. To be even more zoological, its body was laid out like that of *Glandiceps*: one undivided compartment forward (bearing the tentacles) and two divided compartments to the rear. Polyzoans have only two compartments, one in the front and one in the rear. Therefore, according to Harmer, it was a hemichordate, rounding out the scheme of our closest relatives, despite being a minute colonial tentaculate animal rather than a large solitary worm like *Glandiceps*. It is now classified as a pterobranch hemichordate because its expanded tentacle crown looks a little like a pair of wings (think of pterodactyl, the winged reptile of the Mesozoic). This "very unexpected result" (as Harmer puts it) raises a point of fundamental importance: when we are trying to work out the relationships among animals, and therefore by implication our own place in the natural world, should we place more importance on obvious features that are plain to see, such as a wormlike body versus tentacles, or on the ground plan of the body, such as two compartments versus three? M'Intosh hedged his bet, cautiously placing *Cephalodiscus* "near the Polyzoa." He was wrong, and Harmer was right: *Cephalodiscus* is a hemichordate. Gill slits do not lie.

There is an unexpected coda to this unexpected result, which needs a little background to explain. A few groups of animals, such as trilobites and ammonites, become iconic fossils because they were abundant and diverse for many millions of years before becoming completely extinct. One such group is the graptolites, which appeared in the Cambrian period, prospered mightily for nearly 200 million years, but then disappeared completely in the Carboniferous period, about 370 million years ago, never to be seen again in the fossil record. They were elongate animals formed of a strip or a bundle of strips that look like the mark that a soft-lead pencil would make on the rock. When magnified, each strip has a sawtooth appearance showing that it consists of a linear colony of small zooids. In most species the colony was suspended from a float and lived at the surface of the sea, although a few grew on rocks or shells at the bottom. Graptolites are very abundant as fossils, but their soft parts are seldom preserved and their relationship to other groups was correspondingly obscure. The suggestion made in the 1950s by the Russian biologist V. N. Beklemishev that graptolites were pterobranch hemichordates was not widely accepted, partly because it seemed implausible (the colony structure is quite different) and partly, no doubt, because he published his idea in an obscure journal in Russian. The idea eventually took hold, however, and has now become the conventional interpretation. In which case, of course, graptolites never became extinct at all, because pterobranchs are their lineal descendants. They merely retreated from their prominent position on the high seas to take up a more humble existence as rather rare and obscure animals living on the sea floor. In place of the thousands of species of graptolite, only a couple of pterobranchs are known today: *Cephalodiscus* itself and a related form called *Rhadopleura* that the naturalists had actually bottled over two years previously, off Tristan da Cunha, and, unaware of its real nature, included with the Polyzoa collected in the same haul. The expedition had indeed collected living fossils, although this was not made clear until a century later, and rather than lying hidden in the abyss these animals live in shallow water from which Forbes could have collected them, hauling in his dredge by hand from a rowing boat.

There is one further twist in this curious story. Some mutilated scraps were collected off Kerguelen in January 1874 and put in a bottle labeled "Sponge." After the voyage they were first sent to M. S. O. Ridley, who was writing the report on demosponges, and finished up later at the British Museum (Natural History) in a bottle labeled "Hydroid?" Fifty years later they were recognized by one of the museum staff as *Cephalodiscus* and sent on by the Director of the Museum to a specialist, W. G. Ridewood, for further study. The Director of the Museum at the time was Sidney, by then Sir Sidney, Harmer.

Two days later the ship anchored at Port Famine, near the western end of the strait, and steamed from there to Sandy Point (Punta Arenas). By 20 January they were off Monte Dinero, at the eastern end of the strait, and after crossing the Sarmiento Bank headed out to sea, in the Atlantic again, bound for the Falklands.

Station 318: 11 February 1876

42°32′ S, 56°29′ W; Argentine Basin, 2,040 fathoms

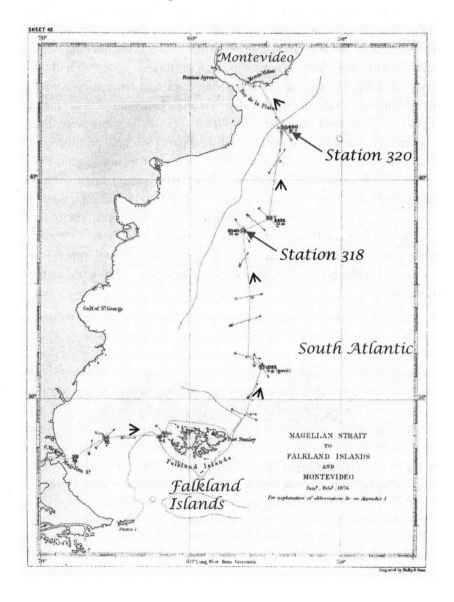

The sea is shallow all the way from the mainland to the Falklands, never more than 200 fathoms and generally a good deal less. The islands were sighted, in fog, on 22 January, and the ship anchored in Port Stanley the following day. Although this was the southern summer, it was bitterly cold and the boats rowed ashore in a hailstorm. Murray notes gloomily that the islands "are a treeless expanse of moorland and bog, and bare and barren rock," and Matkin found the landscape even more dismal than Patagonia. In any event, the naturalists seem to have done very little zoology during their stay, although Moseley discovered some interesting wingless flies that, like those of other isolated islands, have abandoned flight. The ship took in some coal and the crew feasted on mutton, which was very cheap. Moseley and Channer rode out across the quaking bogs to Port Darwin, near Goose Green, in search of the coal beds that had been reported there. The ship took on wooden piles sent from the mainland, there being no wood to be had locally, to make a fence round the cemetery at Port Louis, a short sail up the coast from Stanley, but this helpful gesture ended in tragedy. That evening a dinner party was held on shore, with abundant liquid refreshment, and as the boat was returning Thomas Bush, a popular able seaman, fell overboard, being drunk, according to Matkin. Lieutenant Carpenter very courageously leaped into the freezing water to rescue him, but nearly succumbed himself; Bush was beyond help. Next day the timber was offloaded and the fence erected, while Maclear read the funeral service as Bush was buried, the first grave in the newly refurbished cemetery. *Challenger* left Port Stanley for Montevideo a few days later, on 6 February.

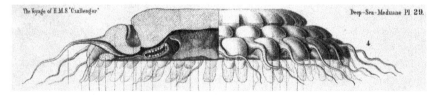

Figure 73 *Atolla* from Station 318
Source: E. Haeckel (1882), Report on Deep-Sea Medusae. Zoology, Part 12, Plate 29. Bound in Reports, vol. 4.

The angler

The water grew deeper as *Challenger* passed across the western fringe of the Argentine Basin, a bowl-shaped abyssal plain some thousand kilometers across, shaped as if some gigantic ball had been pressed into the seabed. The trawl was lowered five days later in 2,000 fathoms, with disappointing results: a few shrimps, a fish, and a deepwater medusa. There were no indisputably benthic animals, like holothurians or ophiuroids, so it probably captured a few animals on its way up or down. The medusa was interesting, though: a coronate scyphozoan, the adult form of the *Stephanoscyphus* polyp that had been found last July attached to manganese nodules, and on many occasions before and since. Such medusae are common in the deep midwater, one of the community of gelatinous predators that also includes ctenophores and siphonophores. They are the most beautiful of animals. Their elegant domed body sways through the water, driven by the soft pulsations of its bell. By comparison, echinoderms are warty creatures, crustaceans clunky, worms mundane, and sponges mere lumps. Even the most lissome fish have bodies that seem stiff and blocky next to the fragile intricate architecture of a medusa. This is a minority opinion, of course. When medusae are abundant enough to be noticed, they are generally represented as a plague. The popular press announces "Attack of the Blob" or "Blight of the Jellyfish," and even reputable science journals find themselves "imagining the jellyfish apocalypse" and warning that "the stinging gelatinous blobs could take over the world's oceans." Blob, indeed. I wonder if any of the journalists who write such stuff have ever seen a living medusa, shimmering in the water.

The elegance of the medusa extends to its killing apparatus, the battery of miniature poison harpoons borne on the tentacles. These masterpieces of precision engineering are far more pleasing than the continual flailing of filter feeders, the grubby habits of deposit feeders, or the suck-and-gulp of fish. They are also very effective: scyphozoan medusae are often the dominant pelagic predators, down in the dark zone. This particular medusa, *Atolla*, along with the usual rim of tentacles, deploys also a single long tentacle, many times longer than the diameter of its bell, which trails behind it as it swims and reacts immediately to contact with any potential prey. The tentacle is a fishing line: *Atolla* is angling for siphonophores.

Station 320: 14 February 1876

37°17′ S, 53°52′ W; mouth of Rio de la Plata, 600 fathoms

The espalier animals

The trawl was lowered once more before reaching Montevideo, off the mouth of the Rio de la Plata. This time it did reach the bottom, and brought up an extraordinary variety of animals, including no fewer than 100 new species. Ten of these were stony corals, unexpected in such deep water. Moseley made a special study of them later, using the same serial sectioning technique that Harmer had used to demonstrate the true nature of *Cephalodiscus*, and with equally surprising results. First he had to decalcify the animals, then embed

Figure 74 *Allopora* from Station 74

Source: H. N. Moseley (1881), Report on Certain Hydrozoan, Alcyonarian and Madreporian Corals. Zoology, Part 7, Plate 14. Bound in Reports, vol. 2.

them in glycerin and gelatin, which permeated their tissues, replacing the original calcium carbonate skeleton and enabling him to make thin sections to work out the three-dimensional structure of the zooids. The remarkable drawings he was able to make in this way have never been surpassed, and proved beyond dispute that these corals were hydrozoans, like the millepores, the fire corals, that Moseley had seen previously in the shallow-water reefs of Tahiti. Most of them grow as a two-dimensional branching network, firmly anchored to a hard substrate such as a rock. They resemble an espaliered tree, stapled to a sunny wall. This is a good body plan for a sessile animal that relies on catching its prey from a stream of water, rather like gorgonians such as the fan corals that live on the walls of channels in shallow-water reefs. The trawl had plowed through a garden of these corals, growing like the animal gardens of the fjords, scooping up both the corals themselves and all the animals living on and around them.

Moseley united these deepwater hydrozoan corals, the stylasterids, with the millepores as a new group of hydrocorals characterized by their stony calcium carbonate skeletons. On the other hand, the two groups have different kinds of tentacle: stylasterids have the usual kind, pointed at the tip, whereas the tentacles of millepores are tipped with a little knob. Either calcification or tentacle structure can be used to divide hydrozoans into groups, so which should it be? This is the problem that arose with *Cephalodiscus*, and will arise whenever we try to classify animals, or for that matter languages or pottery or flint hand-axes. The natural classification is identical with the pattern of descent, but how do we know whether resemblance is due to ancestry rather than the independent acquisition of a similar state? To avoid a long digression, I must brusquely dismiss the last 100 years of phylogenetic theory to say that Moseley was probably wrong. The tentacles have won, and the two groups that he united have now been separated, implying that the habit of building stone walls evolves rather readily.

The following day land was sighted, and the ship anchored at Montevideo, two miles from the shore (she drew too much water to approach closer). Being so far out, they had a miserable time with strong winds and high seas that made it difficult to land boats, and it was intolerably hot. Little was done, apart from an excursion to Buenos Aires. Sublieutenant Swire, dangerously ill from a disease contracted in Tahiti, was invalided home from here on the mail ship. Matkin comments: "His father is, I believe, domestic Chaplain at Windsor Castle, and don't know what he will think when he learns the nature of his disease."

18

Homeward Bound

February–May 1876

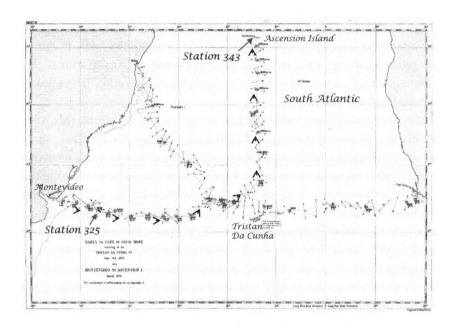

In the early morning of 25 February, in fine weather for a change, *Challenger* steamed out of harbor in the direction of Tristan da Cunha. They trawled that afternoon in very shallow water, only 13 fathoms, and brought up, along with the river mud and many other animals, a bunch of sea pansies, *Renilla*. They don't belong here, really, because they grow only in warm shallow water, but they are nonetheless irresistible. They were not flowers, of course; they were pennatulids, related to the much larger sea pens caught off Madeira early in the voyage, nearly three years earlier. They look rather like a small plant, though, with a stout stem firmly embedded in the sediment, ending in a single, flat, kidney-shaped leaf. Like all pennatulids, the animal consists of a colony of zooids differentiated to perform different functions and giving as a whole the appearance of a unitary individual. The stem itself is a single

Full Fathom 5000. Graham Bell, Oxford University Press. © Graham Bell 2022.
DOI: 10.1093/oso/9780197541579.003.0019

gigantic zooid that anchors the colony. The leaf contains feeding zooids that entangle their prey, usually small animals such as crustaceans, in a sticky goo and then kills them with the lethal anthozoan armament of nematocysts. They also contain specialized zooids that maintain a circulation of water through the animal, and can either inflate it (to catch prey) or deflate it (at low tide, to avoid predators such as hungry nudibranchs). So far, so zoological. But they have another trick: when disturbed, they can generate a bright flash of light. This is generated by the oxidation of a small organic molecule, catalyzed by the enzyme luciferase. So far, so biochemical. Clever biologists, however, have used this trick to engineer the real-time visualization of events in cells or animals, such as nerve conduction, to work out exactly how these events happen. Working out how a sensory signal results in a decision, how a tumor grows, or how gene modification might be used as a therapy have all been greatly assisted by these optical reporters. The two that are used most often are the luciferases from fireflies and from *Renilla*. Nowadays, biochemists can engineer custom-made molecules for different purposes, based on these luciferases. The original engineering, however, was done by the evolution of predator defense in a pennatulid, discovered by zoologists.

Thomson's sailing instructions were quite broad, merely anticipating a possible stop at Ascension Island and expecting that "your return to England might be looked for in the spring of 1876." The Admiralty might have looked for it, but the crew were much more urgently looking forward to it. There is a sort of haste in all their accounts, a loss of detail, which reflects their longing to bring this voyage to an end, and to cover the 7,000 miles from Montevideo to London as quickly as possible. There were a few animals still to come, however.

Station 325: 2 March 1876

36°44′ S, 46°16′ W; northern Argentine Basin, 2,650 fathoms

Away from the coast, the ship moved out into the deep Atlantic again, and began a series of dredges and trawls in deep water that would continue until the end of the voyage. The first was very successful. The trawl came up covered with a network of komoki tubes and containing a mixed bag of animals. Among them was part of the stem of *Rhizocrinus*, the stalked crinoid that Sars had dredged up near the Lofoten Islands in 1864 and that *Challenger* had

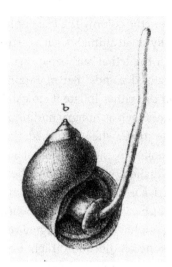

Figure 75 *Pisolamia* from Station 325
Source: R. B. Watson (1886), Report on Scaphopoda and Gastropoda. Zoology, Part 42, Plate 37.
Bound in Reports, vol. 15.

found off Bermuda at the start of the expedition. This was the find that first
suggested that animals thought to be long extinct might still be living in the
depths of the sea. It seemed a good omen, but three of the next four attempts
were fruitless: the trawl carried away on one occasion, losing the best part of
two miles of rope, and came up foul twice. On 2 March they were in luck: the
sailors caught a shark nearly eleven feet long, and the naturalists recovered
the trawl safely, full of animals.

The vampire snail

Among them was the sea pig, *Oneirophanta*, which they had found previously
at many stations in deep water. There were also some small white snails, recorded without comment but in fact attached to the underside of the sea pigs
(as Willemoes-Suhm had observed the previous January, at Zamboanga). In
fact, they were very firmly attached. The front part of most snails consists of
the muscular foot, which it creeps on, and the head, equipped with mouth,
tentacles, and eyes. In this snail, the whole of the front part was thrust through
the skin (leaving the shell outside) to form a sort of fleshy plug from which
protrudes a long proboscis into the interior of the sea pig. This penetrates a

large blood vessel and simply redirects its host's blood to its own stomach, which is why it was named *Pisolamia*, since *lamia* means "vampire" in Greek. Since its food comes predigested, it has no need of intestine or anus, and has only a blind gut. Other parasitic snails go further. One extraordinary animal called *Gasterosiphon* has been reduced to two tubes and a bulb. One tube is plumbed into the main blood vessel of its host—another unhappy holothuroid—and pumps blood to the bulb, buried in the body of the host, where it is used as nutrition for the developing eggs. Once they are mature the eggs pass through the other tube, which goes through the skin of the host, to the outside. That's all. Everything else has been lost—shell, tentacles, eyes, gills, blood vessels, heart, intestine, all gone, so that mere shreds of its remote ancestor, a normal snail, have been retained. Parasites often become highly simplified because they can rely on their host to supply their needs. *Pisolamia* is at least still recognizable as a snail. The extreme simplification of parasites like *Gasterosiphon*, scarcely recognizable even as a separate animal, is a rather startling witness to the inexorable logic of evolution.

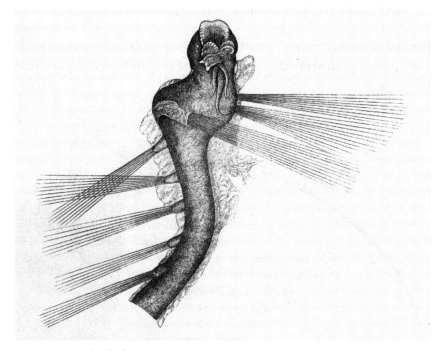

Figure 76 *Buskiella* from Station 325

Source: W. C. M'Intosh (1888), Report on Annelida Polychaeta Zoology, Part 52, Plate 45. Bound in Reports, vol. 24. Also in Narrative, vol. 1, Part 2 Figure 217.

Thomson continued to sail eastward in frustratingly light and variable winds, so the ship made less progress than expected. On 13 March they crossed their outbound path and so completed the circumnavigation, sounding and trawling in the same place that they had visited two and a half years before. They found similar animals, too, including the remarkable pelagic polychete *Buskiella*, which they had last seen in the Sierra Leone Basin. In the net it was little more than a gelatinous lump, but it could be appreciated more clearly when brought into the laboratory and suspended in water in an enamel dish, and nowadays we have pictures and even movies of *Buskiella* in its transparent shift swaying stiffly from side to side as it rows through the dark water with its banks of flattened oar-like bristles. Two days later they were closing Tristan when the wind shifted to dead ahead, and, having little coal left, Thomson decided to sail directly north to Ascension.

Station 343: 27 March 1876

08°03' S, 14°27' W; off Ascension Island, 425 fathoms

The fresh northerly breezes pushed the ship on into the Trades and she made good progress, interrupted only by two dredges that produced nothing

Figure 77 *Pontostratiotes* from Station 332
Source: G. S. Brady (1883), Report on Copepoda. Zoology, Part 23, Plate 45. Bound in Reports, vol. 8. Also in Narrative, vol. 1, Part 2 Figure 313.

from the seabed. The deep tow-nets were more successful and caught a variety of animals, including a small copepod, *Pontostratiotes*, which probably came from the seabed or close above. It belongs to an obscure group, the harpacticoids, which was very seldom collected by the expedition, but is now known to be very abundant in the deep sea. Like the equally minute cumaceans and tabnaids, the harpacticoid copepods are members of the interstitial fauna, living between sand grains and sediment particles in the uppermost layer of the seabed. Even today they are almost unknown; a sample of sediment may contain hundreds of them, most of which belong to species that have never been described. If you would like to name some new species, then a bucketful of ooze will supply them, although you'll need to spend five or six years studying harpacticoid systematics to identify them.

For the next 3,000 miles, from Tristan to the Romanche Fracture Zone, the ship was sailing along the top of the southern part of the Mid-Atlantic Ridge. Ascension was sighted, in heavy rain, on the morning of 27 March, and on the approach to the island the dredge was put over again. This time it was more successful: the dredge itself did not contain much, but a variety of animals were caught in the tangles attached to the net. They had found yet another coral garden.

The free rider

The coral this time was *Metallogorgia*, named for the beautiful metallic sheen of its skeleton, which shows clearly through the thin external layer of tissue. Another animal in the trawl was an ophiuroid with the evocative name of *Ophiocreas oedipus*. Both the coral and the brittle star were new and, being such different kinds of animal, were described separately in different Reports; it was not known until long afterward that they are inseparable companions, always found together. It seems that a small juvenile *Ophiocreas*, scarcely bigger than an *x*, initially settles on a minute young coral consisting of only a dozen or so polyps. As the coral grows and ramifies, the ophiuroid clambers upward, until eventually the mature ophiuroid occupies the crown of the mature coral. The advantage of this arrangement for *Ophiocreas* is very clear: it is raised above the surface of the sediment into a stronger current of water where its extended arms find a richer source of prey. It is also protected against predators by the stinging cells of the coral. The advantage for the coral is not as clear, and perhaps there is none, any more than a bush benefits

Figure 78 *Ophiocreas* from Station 343
Source: T. Lyman (1882), Report on Ophiuroidea. Zoology, Part 14, Plate 31. Bound in Reports, vol. 5.

from the sparrows that shelter on its boughs. It seems likely that *Ophiocreas* is simply a freeloader taking advantage of the platform provided by the coral, like the animals that Willemoes-Suhm noted in the Philippines the previous January. Even so, it has two problems to solve. The first is to avoid being attacked by the stinging cells of its perch; how it does this is unknown, but presumably involves a trick that works only with this species of coral, since it does not seem to use other platforms. The clown fish that lurk among the tentacles of sea anemones in shallow-water reefs have evolved a similar immunity. The second problem is to police its exclusive tenancy, since any coral is occupied by only a single *Ophiocreas*. How it does this is also unknown; perhaps it simply eats any juvenile that tries to settle. These puzzles must be left for future zoologists, but the fact of the association between such dissimilar partners reminds us that the bag of animals brought up by the trawl are not merely new and unknown species but also members of a complex community stitched together by myriad unexpected interactions.

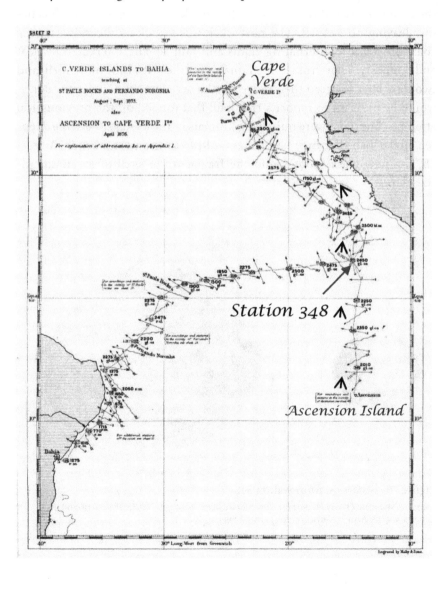

Station 348: 9 April 1876

03°10′ N, 15°51′ W; Sierra Leone Basin, 2,450 fathoms

In the afternoon they stopped in Clarence Bay, the only useful anchorage on an island known as "Sailors' Hell" and consisting, according to Matkin, of little more than mountain, cinders, and sand. There was no fresh water and no produce of any kind except turtles, which they ate at most meals, flesh and soup and eggs. It was, however, an important strategic base for the Royal Navy, and the ship was able to take on coal and provisions before leaving on 3 April. The next week was vexatious, steaming up the coast of Africa in alternating calms and rain squalls, enervated by heat and humidity. They crossed the Equator for the sixth time on 7 April, and two days later lowered the dredge for the very last time in deep water. It would be pleasant to report a rich haul that topped off the previous 150 attempts, but there were only three animals in the dredge: the same deep-sea angler fish that they had caught a little to the west, back in August 1873, a new glass sponge, and the fragile white shell of a thumbnail-size clam.

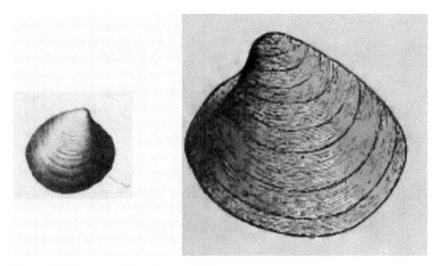

Figure 79 *Vesicomya* from Station 348

Source: E. A. Smith (1885), Report on Lamellibranchiata. Zoology, Part 35, Plate 6. Bound in Reports, vol. 13. Also in Narrative, vol. 1, Part 2 Figure 210.

The last clam

The clam, *Vesicomya*, was the most interesting of the three. It belongs to a group with two main branches. One includes the giant white clams, a foot long, found at hydrothermal vents. They have bloated gills stuffed with symbiotic bacteria that are able to oxidize the hydrogen sulfide dissolved in the hot vent water, and they live by farming these bacteria rather than by filtering plankton like other clams. As you might expect (use it or lose it), their gut has become reduced to a rudiment. The other group—to which the animal dredged up by *Challenger* belongs—are much smaller and more broadly distributed. Could the naturalists have finally succeeded, at the last throw of the dice, in sampling a hydrothermal vent and bringing up one of its inhabitants? Some authorities assert that all species in the group live in sulfide-rich environments and have symbiotic bacteria—there are even published photographs of bacteria in their gills. Moreover, *Challenger* was trawling only a little to the north of the Romanche Fracture Zone, where the crest of the Mid-Atlantic Ridge swings west, in an area where vents are known to occur. It is tempting to declare a fitting finale to the expedition, another major discovery, however little appreciated it may have been at the time. Sadly, there are good reasons to be skeptical. The gills of *Vesicomya* are not notably enlarged, and lack the specialized filaments that harbor bacteria in the big vent clams. The gut seems to be functional, and they have both inhalant and exhalant siphons (although both are short). They may not even be restricted to sulfide-rich environments, but range more broadly across the sediments of the abyssal plain. Their ecology is obscure enough, and their taxonomy is certainly confused enough, to hold out the hope that some species at least illustrate an intermediate stage in the evolution of their spectacular relatives. On the whole, though, I doubt that the naturalists had finally succeeded in clipping the corner of a vent field; but they certainly raised a question that has not yet been conclusively answered.

Homecoming

After calling at Cape Verde they made good progress at first toward the Azores, but then found headwinds that blew them to the west. Thomson was forced to call at Vigo, on the Spanish coast just north of Portugal, to coal.

They passed the Channel Fleet on their way in and took in enough coal to see them through to England, working till midnight to fuel the ship for its final run. The weather was still rough, but the ship weathered Cape Finisterre and thankfully made a swift passage across the Bay of Biscay, sighting the Ushant light on the evening of 23 May. The morning after was foggy, but Start Point on the coast of England was sighted, at long, long last, and they proceeded up Channel in clear cool weather and anchored at Spithead that evening. Home.

Wyville Thomson and the rest of the scientists promptly went on leave, or at any rate left. Wyville Thomson himself traveled to Edinburgh to be reunited with his wife Jane, who had last seen him three and a half years ago. No doubt she had developed her own routine during his long absence, and would have found this interrupted by the return of her husband. Both would have aged a little, too, and perhaps noticed this, or thought the other did. Absence may make the heart grow fonder, but a long absence makes a stranger out of a lover, and they would have had to cope with this, as all naval families did in those days of long voyages. Murray and Buchanan also returned to Edinburgh, Moseley to Oxford; none were married at the time. With that, the company that, with Willemoes-Suhm, had formed the scientific staff of the *Challenger* was broken up, and never met again as a group, as far as I know, although Wyville Thomson and Murray continued to be closely associated for several years during the initial preparation of the Reports.

Officers and crew had shore leave, and the ship swarmed with welcomers—family and friends, together with more dubious figures eager to buy or sell. But the voyage was not quite finished yet. Early the next morning the ship moved out in cold, wet weather (welcome to England!) and anchored at Sheerness on the Queen's birthday. The ship was thronged with tourists, arriving by special trains from London, who inspected the dredging apparatus with wonder. The officers laid on a lunch for the visiting dignitaries (Campbell came too). It went on for a week, until the Admiralty ordered the ship down to Chatham for paying off on 6 June. One of the officers needed no paying off. Sublieutenant Henry C. E. Harston, who had joined the ship in the Cape Verde Islands on the outward voyage in 1873 (where he had enjoyed the hospitality of the Consul), excused himself from dinner and stayed in his cabin, where he committed suicide by drinking a large amount of chloral hydrate. He left a letter addressed to the surgeon apologizing for the trouble he had given. His parents were Henry Harston, himself a naval captain, and Alicia, the daughter of a clergyman; no apologies would assuage their sorrow.

The crew were paid off on Sunday, 11 June, and the crew granted a week's leave. Some had kept their money in the ship's savings bank, but many had spent it on souvenirs or drink or whores. The storerooms were cleared and the accounts closed. The men walked away, later to re-enlist or to take their discharge from the Navy. The ship shifted restlessly against the quay, as ships do. The voyage was over.

PART 3

AFTER

19

What Happened to the Ship

Both Forbes and Murray, the two poles of the expedition, were amateurs. The expedition itself was amateurish, using untried, homemade equipment that had been installed into a halfheartedly converted survey ship. The Admiralty was shocked at the cost of the expedition (about £200,000 at the time) and declined to repeat the experiment. It was certainly unsentimental about *Challenger*, and planned to convert her into a training ship for boys. Even this failed, and her eventual fate was to be a receiving hulk on the Medway, until she was broken up to salvage the copper sheathing of her hull in 1921. A few bits and pieces may have been preserved: Buchanan installed his old chemistry bench in his research vessel *The Ark* at Granton, and perhaps other crew members took souvenirs. Otherwise nothing of the ship now remains except her figurehead—a knight in armor with a rather luxuriant mustache—which is preserved at the National Oceanography Centre in Southampton.

There were a dozen other deep-sea expeditions in the years before the First World War, mostly German and American ships, with contributions from Russia and Norway, including Nansen's Arctic voyage in *Fram*. They have continued down to the present day, with steadily improving ships and gear. All of them do the same sort of things as *Challenger*, although they undoubtedly do everything much better. Manned submarines capable of surviving in the deep sea have been constructed, beginning with the bathyscaphe *Trieste* in 1960, and have enabled a few courageous explorers to visit the regions that *Challenger* could sense only dimly and indirectly, but the scientific results have been quite meager. A more fruitful advance, well beyond anything that the *Challenger* was capable of, has come from the deployment of cameras mounted on remotely operated submersibles. The animals that *Challenger* fished up dead or dying, often mutilated or broken into fragments, can now be seen alive, intact, and going about their usual business. The earliest models date from the 1950s, but it is only quite recently that advanced imaging technologies have produced dramatic pictures of animals that we knew previously only from preserved specimens in bottles. Only the telescope and

Full Fathom 5000. Graham Bell, Oxford University Press. © Graham Bell 2022.
DOI: 10.1093/oso/9780197541579.003.0020

Figure 80 Hulk of HMS *Challenger* at Chatham
Source: Narrative, vol. 1, Part 2 Tailpiece, p. 947.

the microscope surpass these deep-sea cameras in making visible a whole new world that was previously hidden from sight. There are even live feeds to several of them, so you can watch what the naturalists on board *Challenger* could only guess at, the animals that swam or crawled two or three miles below the keel of their ship.

20

What Happened to the People

The return of the *Challenger* was Wyville Thomson's moment of triumph. He had led an expedition that had dredged the deep sea around the globe and brought home thousands of specimens to prove it. He could justly claim to have discovered a new world of animals, never before seen by human eyes. There had been a ceremonial lunch at Sheerness for the Lords of the Admiralty and the Fellows of the Royal Society, to welcome back the heroic band of scientists who had plumbed the world's oceans for the last three and a half years. He must have basked in the admiration of the social and scientific world.

The Death of Another Naturalist

Wyville Thomson had a plan for dealing with the enormous collection of specimens that had been amassed by the expedition. It had been prompted by a letter from the Admiralty to Captain Thomson the previous July, asking for a report from Wyville Thomson describing how he proposed to deal with the specimens collected by the expedition, including a budget for the work. Wyville Thomson replied that the collection would first be sorted into groups at Edinburgh and then distributed to specialists at different museums for naming and description, with the members of the expedition having first choice of groups, as was only fair. This would cost £2,800 in the first year and £4,000 per year thereafter. The naming and description would take about three years, and once this had been done a complete series would be lodged at the British Museum. Material from intermediate ports had already been sent to Edinburgh during the voyage, and the cases brought back by the ship were on their way, to be stored in houses owned by the Edinburgh University. Wyville Thomson had anticipated, in the letter he wrote in Valparaiso, that the specimens would all be returned within three years, to be sent on to the British Museum, and that all the reports would be published within five years.

Full Fathom 5000. Graham Bell, Oxford University Press. © Graham Bell 2022.
DOI: 10.1093/oso/9780197541579.003.0021

Three years later, no specimens were forthcoming and no report had yet been published. Wyville Thomson had not been idle by any means. He had written *The Atlantic*, in two volumes, to describe the Atlantic sections of the voyage, and was preparing a general introduction to the reports themselves. But he was evidently not a good administrator. He understood that the specialists who had agreed to deal with the specimens were busy men, and refrained from badgering them to send in their copy. He was distressed by the very public controversy about his handling of the collections, which I shall mention in the next chapter. He was a poor accountant who failed to keep many of the receipts for expenditure that the Treasury demanded. He was also unwell. He admitted to feeling "not very well" in 1878 and suffered an attack of paralysis (presumably a severe stroke) in June 1879. By now he was clearly a sick man. He retired from the Chair at Edinburgh in October 1881, and soon afterward resigned as Director of the Challenger Expedition Committee. He died in March 1882, in the house in Linlithgow where he had been born. He was only 52 years old.

Murray then took over. He had a very different personality. He was quite willing to badger busy men into fulfilling their obligations. He did not care a rap for the eminent scientists whose criticisms distressed Wyville Thomson. He kept careful accounts and could not be browbeaten by Treasury officials. The first Reports were published in 1880, with Wyville Thomson's introduction, and 82 zoological reports appeared in the next decade, with Huxley's description of *Spirula* closing the series a few years later, in 1895. Murray also wrote (with Alphonse Renard of Ghent) the Report on Deep-Sea Deposits, the field that had become his own specialty. Together with a few accounts of botanical collections, these reports fill the 50 hefty volumes that were distributed, free of charge, to major universities and museums around the globe. They describe a new world of animals: nearly 2,000 species from below 500 fathoms, three-quarters of them never seen before. It seems heartless, even after so many years have passed, to claim that the final success of the expedition depended on the early death of its director. Perhaps the reports would have appeared eventually, spurred on by the mounting impatience of the scientific community and the grumbling of the Treasury. There is no doubt, though, that credit for the annual flood of reports between 1880 and 1889 belongs to the energy and perseverance of John Murray.

The Challenger Medal

The authors of the reports worked for glory and were not paid, except for a small sum to cover expenses. Once the series of reports was complete, however, Murray decided that they needed a little more recognition and struck a medal to reward them. The Challenger Medal is a handsome bronze disk three inches in diameter. It shows Neptune (to symbolize the sea) brandishing a trident, above the head of Athena (to symbolize wisdom), with two mermaids below holding a strategically placed ribbon bearing the message "Voyage of HMS Challenger 1872–76." The other side shows a knight in armor throwing down a gauntlet as challenge to the sea, with a message commemorating the report of the expedition. It was awarded to everyone, or almost everyone, closely associated with the expedition, plus a few of Murray's friends whom he thought merited it for one reason or another, so the list of recipients was more or less the entire cast of the show. The naturalists all got one, of course, even Willemoes-Suhm, one of the few posthumous awards: the medal was sent to his mother. The officers and petty officers all got one too, from captain to carpenter, although the list got a little ragged toward the bottom. There were even two awards to other ranks—Frederick Pearcey, Domestic Third Class (but he had assisted in the chemistry laboratory and was later employed at the Challenger Office) and Richard Wyatt, Writer Third Class (I don't know why). Almost all the authors of the reports received a medal, at least those who were still alive in 1895, although Edward Miers (crabs) seems to have been overlooked; he was a curator at the British Museum, but I don't think that Murray was being vindictive. The Royal Society received a collective medal for teamwork.

John Murray

Murray did not award himself a medal. He scarcely needed to: honors were showered upon him, from election as Fellow of the Royal Society in 1896 and appointment as Knight Commander of the Bath in 1898, to the Cuvier Medal of the Institut de France and the Humboldt Medal of the Gesellschaft für Erdkunde of Berlin. He responded to acclaim by becoming more active than ever. He founded the first Scottish marine biology station at Granton, on the Forth Estuary, complete with a ruined abbey to accommodate distinguished visitors. A year later he founded the second one, at Millport on the west coast. He put his *Challenger* skills to use in sounding the Scottish lochs: with three paid assistants at the Challenger Office he completed bathymetric surveys of all 500-odd lochs within 10 years. At the age of 67 he mounted (and funded) an expedition with the Norwegian Johan Hjort to explore the Faroe Channel and the North Atlantic, and published *The Depths of the Sea* in 1912. But it was the coral reefs that made his fortune.

Murray had thought long and hard about the formation of coral reefs. He was well aware of Darwin's work on the subject, of course, but unlike the rest of the world he thought it possible that Darwin might be wrong. A coral atoll was formed, according to Darwin, when a volcanic cone slowly sank, with the upward growth of the coral keeping pace. Murray imagined instead that the globigerina ooze that mantled sunken volcanos would slowly accumulate until it was near enough to the surface for coral polyps to grow, forming an island that would become a ring-shaped atoll when the coral in the middle died from lack of food. It would have been embarrassing to disagree in public with Darwin, however, and he may have feared for the continued funding of the reports had his heterodox ideas been vigorously propounded. At all events, he held his peace for a long time, but these ideas led him to become interested in Christmas Island, which lies at the edge of the Java Trench about 200 miles southwest of Java. It had not been visited by *Challenger*, so Murray had no samples from it, but being Murray just asked William Wharton, Richards's successor as Hydrographer at the Admiralty, if he could arrange for some to be collected. He could; by an odd chance, Maclear, Murray's former shipmate, was about to pass by in command of the survey vessel *Flying Fish*, and gladly stopped off to take a look at the island. He found that it was covered in dense forest and fringed with low cliffs that made the shore almost unapproachable. It was of no naval interest, but Maclear managed, with some difficulty, to land on a small beach and collect some pieces of coral rock. When these finally turned up in Edinburgh they turned out to contain nuggets of pure calcium phosphate. This was a mildly interesting scientific puzzle, since it was not clear how these nuggets were formed. But Murray also knew all about the use of inorganic fertilizers, which had recently begun to be widely used in Britain, and guessed that the commercial implications might be more than mildly interesting. Murray went back to Wharton and asked for a proper survey of the island, including rock collections from the interior. Wharton may have sighed, but sent the signal anyway, which, by a truly preposterous coincidence, reached yet another old *Challenger* hand,

Pelham Aldrich, now in command of *Egeria*. Aldrich anchored at the beach that Maclear had found, and a shore party led by a couple of enterprising lieutenants managed to scramble over the cliffs into the interior. There was even a naturalist with them: J. J. Lister, the nephew of the surgeon Joseph Lister, who dutifully collected the few animals he could find. Samples of coral rock from the interior were duly dispatched to London, where they turned out to contain nuggets of phosphate similar to the original samples. Murray guessed that there must be high-grade phosphate deposits on Christmas Island, and he was right. In fact there was about 12 million tons of the stuff, which would yield a clear profit, landed in London, of £2 per ton.

Murray's course of action was clear, at least in those days: he asked the government to annex the island so that he could exploit this windfall. (Actually, he asked the Duke of Argyll, who besides being a powerful politician was the father of Sublieutenant George Campbell of the *Challenger*.) After some hesitation—they didn't want to start a war for Murray's benefit, after all—the government agreed to do so, and sent HMS *Imperieuse*, Captain William May, to take possession of the island. At this point, Murray's plans hit a snag in the shape of the Clunies Ross family, which owned coconut plantations on Keeling Island and wanted to set up others on Christmas Island. They managed to baffle Murray for several years, until a joint stock company was set up and several hundred Chinese laborers descended on this lonely spot in 1899 to begin mining. Before long the phosphate ships were arriving in London, the money rolled in, and John Murray, now Sir John, became very rich.

Murray was in any case quite well off, because in 1899 he had married Isabel Henderson, whose father owned the Anchor Shipping Line. The couple moved into a large house in Granton, which they named Challenger Lodge, with Murray's collections just over the road in Medusa Lodge. They lived here in comfort for many years, while his wife bore their two sons and three daughters. Murray surveyed the lochs, cruised in the Atlantic, collected honors, and in 1900, at the age of 67, sailed halfway round the world to visit his mine on Christmas Island.

The Other Scientists

W. B. Carpenter, who had been so instrumental in planning the expedition, did not get a medal either, for he died in 1885. Like Wallace, he was convinced by Darwin's theory of evolution by natural selection, but resisted its application to human thought and spirituality. Not that he was an advocate of spiritualism, quite the contrary, although he thought that telepathy might be possible. He was fond of vapor baths, a sort of domestic Turkish bath, enclosed by a curtain, which enveloped the body in steam. One day the oil lamp used for illumination caught the curtain on fire, and Carpenter was so severely burned that he died a few hours later.

Buchanan and Moseley both received medals, of course, but their careers after the return of the *Challenger* were very different. Buchanan appears at first sight to have been a rather cold and cynical man, although his obituary explains that this was a false impression given by his shyness and wincing sensitivity to criticism; certainly, this type of personality is not uncommon. He remained associated with the Challenger Office for a time, took part in fitting out the Grafton station, and cruised along the west coast of Scotland in his stream yacht *Mallard* (not a poor man, then). In 1889 he moved to Christ's College, Cambridge, in the relatively lowly position of lecturer in geography and seems to have fitted in well there, but later went back to sea, laying cable off the west coast of Africa, and collaborating with Prince Albert I of Monaco, who was keen on oceanography. He survived the expedition by half a century, and died in 1926 at the age of 81.

Moseley was a very different personality, a bluff, outgoing man who made friends easily and kept them for long. Perhaps that is why he could not resist an invitation by Wallis Nash, an English lawyer, to visit the Oregon territory with a small party the year after *Challenger* returned. It was no small matter in those days to travel to Oregon, but Moseley went, and when he returned wrote a book about Oregon, a very dull book that includes earnest advice to potential immigrants. Nash was serious about Oregon and later went to live there, to build a railroad and to found an agricultural college that later became the Oregon State University (whose most famous alumnus was Linus Pauling). I have no idea what Moseley's motives were for this random adventure, but his career certainly bloomed after his return. He was appointed Linacre Professor at Oxford in 1881 and married the same year. He plunged into teaching and zoological research, and in particular identified *Peripatus*, the peculiar wormlike animal that he and Willemoes-Suhm had found in farmyards at Cape Town, as the sister group of all the arthropods. His broad interests included anthropology, and he was instrumental in securing the vast Pitt-Rivers collection for the university and arranging its exhibits. (You can find it now at the rear of the University Natural History Museum; its arrangement is little changed from Moseley's day, and can be viewed as a charming survival or an archaic jumble, as you wish.) He was launched on a brilliant academic career, with a growing family and no money worries. But he then began to suffer from migraines and depression, and in June 1887 his health broke down and he was nursed by his wife for the last four years of his life. An attack of bronchitis carried him off in November 1891; he was 47 years old.

Officers and Men

Sir George Richards, Hydrographer to the Admiralty, retired in 1874 and promptly became managing director of the Telegraph Construction and Maintenance Company. There is no suggestion (I have looked) that he was associated with the company during the planning of the voyage, but the door certainly revolved quite rapidly. He died full of honors—Fellow of the Royal Society and Admiral must be an unusual brace—in 1896.

Sir George Nares went on his Arctic expedition and penetrated in the steam sloop *Alert* as far as the northern tip of Ellesmere Island. He lost several men from scurvy, and the subsequent inquiry was critical. He was sent to survey the Magellan Strait, but after little more than a year was superseded (by Maclear of all people) to be appointed to the Harbour Commission, with a knighthood for compensation. Nares was a sound seaman and a well-liked captain, but he never saw action and twice fumbled the pass, with desertions from *Challenger* and scurvy on *Alert*, and his reputation has never been as high as his geniality might deserve. He died in 1915; one of his sons followed in Richards's footsteps and became Assistant Hydrographer to the navy. Frank Tourle Thomson, Nares's successor on the *Challenger*, went on to command a paddle steamer at Portsmouth and died in 1885.

Most of the other officers had conventional naval careers, either as sea officers or as surveyors. Few saw action, as this was a period of general peace, at least compared with the disastrous century that followed. Aldrich served with Nares on *Alert* and led the Western Sledge Party through bitter spring weather and deep snow to northern Ellesmere Island, only a few hundred miles from the North Pole. The northernmost point of land in North America (perhaps, especially if you rule out Greenland) was named Cape Aldrich in his honor. He eventually took up a shore appointment as Superintendent of Portsmouth Dockyard. Bromley rose through the ranks to command HMS *Hood*, not the famous one, but its pre-dreadnought precursor that patrolled the Mediterranean. Tizard, Carpenter, Channer, and Havergal chose surveying and hydrography. Campbell is difficult to pin down; with his racy *Log-book* and *Journal*, not to mention being the son of a duke, he seemed destined for a glittering career, but he was still on the Navy List as lieutenant in 1885, by which time the glitter would have worn off. He died in 1915, just before his only son Ivar was killed in action against Ottoman troops at Sheik Sa'ad in Mesopotamia. Two of the officers chose other careers. Henry Sloggett resigned his commission and became a medical student at Glasgow. He later emigrated to Hawaii and became President of the Medical Society of Hawaii. George Bethell was even more imaginative and was elected MP for Holderness, in Yorkshire, in the general election of 1885. He lacked the political gift, however, and was deselected (as we would say now) by the local Conservative Association, over an argument about government policy in South Africa, and did not stand in the next election.

The seamen received no medals, of course. Eighty of them were scattered around the world, wherever they had deserted from the ship. Those who returned to Sheerness were assigned to other ships or left the service, as Matkin did. He returned from the ship to his native town of Oakham but soon moved to London and became a junior clerk in the Civil Service. He married Mary Swift, whom he had met in Oakham, and lived in Croydon with her and their five sons for the next decade. He then abruptly retired at the surprisingly early age of 41 and moved back to Oakham with his family, although how he was able to manage financially is far from clear. At some point he separated from his wife and moved back to London, where he seems to have been killed in a street accident, perhaps run down by a motorbike, in 1927.

The End

Murray held progressive views as well as being rich. He not only bought one of the newfangled automobiles but even taught his daughter Rhoda to drive it. He was with her in the car one day in March 1914 when she attempted to change gear on a straight and level stretch of road, put the car into a skid, and ended up overturned in the ditch. She was knocked unconscious; John Murray was killed instantly. His death closed the account of the voyage, just a few months before the end of the long Victorian era itself. A few people linked with the expedition lived on for many years, like Matkin; James Chumley, Murray's secretary in the Challenger Office, lived until 1948 and may have been the last survivor. The mission of the expedition, however, had been accomplished long before, as the Reports were read and digested.

Some of the ablest and most prominent scientists of the day had been involved more or less directly in the *Challenger* project. They had prepared the ground, petitioned the Admiralty, and sailed around the world, sounding and dredging along the way. It had taken little more than 20 years for their combined efforts to drive the project forward from Edward Forbes's obsession to the return of the *Challenger*. When the ship at last dropped anchor at Sheerness, their grip might have slackened a little, perhaps, now that the main business of the expedition more or less over. Only John Murray was prepared for another 10 years of constant effort. In the end, the lasting imprint on history that the voyage made depended on the pertinacity of a young Scots Canadian who had initially played only a minor role in the project. He has been forgotten in his adopted country and has never been honored in the land where he was born. It was Murray nevertheless who secured the reputation of the *Challenger* voyage as the first systematic exploration of the deep sea, and laid the basis for all that we have subsequently discovered about half the surface of the earth and the strange animals that live there.

21

What Happened to the Animals

Challenger had brought back bottles and tables: bottles of specimens and tables of data. The sludge dumped onto the dredging platform had been transformed into cleaned animals preserved in hermetically sealed jars. There were about 6,000 glass jars and tubes altogether, together with tin cases and casks, containing in all something like 100,000 individual specimens. Once the celebrations were over, Wyville Thomson and Murray were ready to begin work. The collections moved out of Edinburgh University and into the Challenger Office, at 32 Queen Street, in a very Edinburgh terrace of three-story, flat-faced, stone houses overlooking the public gardens. From here, they would be distributed to specialists in Britain, Europe, and the United States who would each contribute a report on a group until every animal collected by the expedition had been identified, named, described, and cataloged.

A Takeover Bid

The British Museum disagreed. Or, rather, Richard Owen disagreed. Owen was then the Superintendent of the Department of Natural History, and a very powerful figure in British science, if not a universally admired one; he was a vain, prickly man of great accomplishments as a comparative anatomist who is now chiefly remembered as one of the principal opponents of Darwin's theory of evolution by natural selection. Owen had received a memorandum about the dispersal of the *Challenger* collections from Albert Gunther, Keeper of Zoology, on 6 June. I suspect that he asked Gunther to write it, and probably dictated its contents, because with suspicious promptitude he wrote the next day to the Museum trustees suggesting—demanding—a very different plan for the *Challenger* collection. His views were communicated to the Treasury on 12 July. Wyville Thomson could scarcely have suspected that, even as he was returning to Edinburgh and his former

Full Fathom 5000. Graham Bell, Oxford University Press. © Graham Bell 2022.
DOI: 10.1093/oso/9780197541579.003.0022

ship was being paid off, a letter was being written in London that threatened to undercut everything that he had achieved.

> In no other country has so little care been taken in securing the scientific collections made at the nation's expense for the National Museum. . . . Professor W. Thomson recommended the Admiralty to deposit the collections (said to consist of some 600 cases) with a professor in Edinburgh, who has no connexion with the government, whilst part of the collections were distributed among other private persons: all communication with the British Museum was carefully avoided.

The collection should instead be promptly deposited with the British Museum, whose staff would take care of it. Owen had played no part in the instigation, planning, or funding of the expedition, and no member of the museum staff had participated in it or contributed to its success. He now proposed to take exclusive possession of all that it had brought back and garner the fame and prestige that was at that time attached to the first description of new species. He was quite right (although one can see why he was not universally loved). Any collection in private hands is almost certain to deteriorate and will eventually be frittered away, lost, or destroyed, without ever adding to the permanent body of scientific knowledge. To be useful, specimens must be accessioned into the permanent collection of a public museum, where they will be cataloged, labeled, stored, curated, and made available to scholars and scientists, in perpetuity. Owen might have had in mind the dreadful precedent of the *Erebus* and *Terror* expedition to the Antarctic. James Clark Ross amassed an enormous collection of marine animals that would have been a worthy forerunner of the *Challenger* collection, but he never quite got round to working on them, and the remains were found after his death, a heap of broken and empty bottles in an outhouse.

The Treasury consulted the Admiralty, who were perfectly willing to have the Treasury decide what to do with the collection. The Treasury agreed with Owen, and wrote to Wyville Thomson in late September to inform him that "the staff of the Natural History Department at the British Museum is ready to undertake, as part of their ordinary official duties, the work of preserving, arranging, naming and describing these collections"—that is, in plain language, to take all the credit. (Whether or not the staff of the British Museum actually had the firepower to deal adequately with all, or even most, of the groups of animals in the *Challenger* collections might be doubted, but the question was not raised.) Wyville Thomson could make an alternative proposal, but in the end the Treasury would decide. Before coming to a decision, however, the Treasury asked for advice from the Council of the Royal Society, quite properly, since this was the body that had drawn up the original plans for the expedition. At this point, Owen's takeover bid began to falter. The Council had endorsed Wyville Thomson's original proposal, and Owen had few friends there, least of all the President: Huxley had been Owen's opponent in the increasingly bitter arguments over evolution that had been going on since the publication of the *Origin of Species* in 1859. Nor was Huxley impressed by the insinuation that Wyville Thomson intended to keep the *Challenger* material permanently in Edinburgh as the basis for a rival national museum. (The new building of the Edinburgh Museum of Science and Art had opened in 1866.) The council recommended, predictably, that Wyville Thomson's proposals for the description of the collections should be approved. The Treasury, faced with conflicting advice from the national academy and the national museum, inevitably took the side of the academy; they had asked for the expedition, after all, and probably knew the Fellows personally, whereas they would certainly not know the obscure taxonomists of the museum. They even agreed to provide funding.

Not everyone was pleased when all this became public knowledge the following year. P. Martin Duncan, the President of the Geological Society, was incensed. "It is absolutely necessary," he wrote in a published article, "to thoroughly open up the question of Sir Wyville Thomson's administration in its widest sense." He listed the people he expected to receive the *Challenger* material, fairly accurately, and was angry that many of the most important groups were to be sent abroad for description, when competent British naturalists had been ignored. He was particularly angry that no one at the British Museum, except Woodward and Gunther, had been invited to contribute. He anticipates darkly that "when the Government bring forward the motion of supply they will be informed that their liberality has been far in excess of the requirements of the case." No doubt the Treasury was grateful for his warning, but it was too late. All the material had been stored at Edinburgh, and much of it had already been sent out. The land animals had already been sent to the British Museum, since they were peripheral to the main purpose of the expedition, and the plants to Hooker at Kew. The great bulk of the collections, however, comprising all the marine animals and the sea-floor deposits, were held at Edinburgh and distributed from there. Duncan complained that it had been sent abroad to two American and four German institutions, but he was wide of the mark: material had also been sent, or was shortly to be sent, to Uppsala, Odessa, Copenhagen, Graz, Oslo, Utrecht, Leiden, and Ghent. About one-third of all the Reports that were eventually published were written by foreign scientists, who handled some of the most important deep-sea groups, including all the sponges, the alcyonarians and pennatulids, the siphonophores, the pycnogonids, many of the crustaceans, and most of the echinoderms. The Royal Society would have been well aware of the implications of its recommendation, but the Society, then as now, regarded science as an international rather than a national enterprise, and understood that the value of foreign collaborations outweighed the patriotic (and perhaps somewhat self-interested) objections of Owen and Duncan.

The Reports

The Reports are the final distillation of the expedition. They cover all of its activities: the physical and chemical measurements, the sea-floor deposits, some plant collections, and above all the animals. The animals had been sorted from the mud and rubble on the dredging platform, preserved in glass bottles in the laboratory on board, shipped to Edinburgh, and distributed to specialists; now they were transformed into drawings and diagnoses and redistributed around the world as printed volumes. These volumes are also the main reason for the enduring influence of the expedition. Some were relatively trivial (Entozoa) or peripheral to the main object of the expedition (Human Skeletons) or concerned primarily shallow-water groups (Stomatopoda). Others were massive tomes dedicated to the principal members of the deep-sea community, such as holothuroids or glass sponges, and written by heavyweights such as G. O. Sars or Alexander Agassiz. To understand better why the Reports were so influential, let us turn to something in between, such as the report on pycnogonids written by the Dutch zoologist P. P. C. Hoek. Pycnogonids are the extraordinary asterisk-shaped animals, all legs and no body, that I described from Station 50, out from Halifax in the early days of the voyage.

Hoek was a rather junior scientist at the time, working as an assistant at the Zoötomisch Laboratorium in Leiden and teaching in the local high schools. (His parents clearly had high hopes for him: his initials stand for Paulus

Peronius Cato.) Wyville Thomson offered him the pycnogonid material in June 1877, and this was not an ill-advised decision to ship the specimens to an unknown foreigner: Hoek had been recommended by Charles Darwin. He presumably agreed, although Wyville Thomson lost his letter (typical) and had to ask for a duplicate. He would have received the specimens in early 1878, and had completed his report in a very timely fashion by November 1880. It was published the following year and praised highly by Darwin; just as impressively, Alphonse Milne-Edwards (who despite his very English surname was an eminent French zoologist) called it the most important work yet done on the group. This shows the Challenger Office at its most efficient: publishing an authoritative report within three years of dispatching the specimens.

What was in the report? It begins with an historical survey of previous work on the group, and Hoek is confident enough to criticize earlier authors quite severely, including John Goodsir, who had dredged with Forbes. He next gives an overall description of the *Challenger* material, which is quite limited, amounting to only 120 specimens from 26 stations. He points out that this is by no means an exclusively deep-sea group, with most of the specimens coming from relatively shallow dredges in 500 fathoms or less. The species from deeper water tend to have a rather similar appearance, however, with long, slender legs and a smooth body. Hoek gives a catalog of the species that had been described previously, and then names and describes the specimens collected by the expedition, arranging them into 36 species, of which all but three are new to science. There are then two appendixes, a brief account of a later cruise and a more extensive account of the anatomy and embryology of the group. The entire Report is just short of 200 pages, illustrated by a series of plates with the careful and exact drawings typical of those (largely) prephotography times.

All of this sounds worthy but dull, which is exactly what it is. You are unlikely to choose Hoek for bedtime reading unless you happen to be unhealthily fond of pycnogonids. But it does show why the lengthening file of reports made such a strong impression. Most of them were authoritative; they named a raft of new species; they described a new community of deep-sea animals; and they had detailed (and often gorgeous) illustrations. If you wanted to discover what was known about any group of marine animals, you would find an answer somewhere in the 15 feet of *Challenger* reports, filling a shelf in the local museum or university library of most major centers throughout the world.

Physics, Chemistry, and Politics

A great deal of time on the voyage had been taken up with measurements of the temperature, salinity, and specific gravity of seawater, and these too were duly reported. The endless records of temperature and salinity from water samples laboriously hauled up on the sounding rope had been reduced to neat columns of figures. The obvious person to write up these reports was Buchanan, who had compiled them assiduously on the *Challenger*, in his cramped chemistry laboratory, for the whole of the voyage. But he made enemies too easily. Wyville Thomson wrote rather cold-bloodedly to Tizard in 1879:

> It is fortunate that Buchanan has already written what I think is ample about the Hydrometers, for after representations he made to the Treasury, they sent me very sharp orders that he was not to be employed again.

He never was; his sole contribution was the brief report on specific gravity that appeared in 1884. It was of no great value. He had made a crucial contribution to two of the major discoveries of the expedition: the composition of the nodules strewn so thickly over tracts of the Pacific Ocean, and the inorganic nature of the illusory *Bathybius*. He could have written up either, or both, and might now be remembered as the father figure of marine mining or the courageous chemist who defied Huxley and Haeckel, but in fact he has been entirely forgotten. It really does not pay to be rude to the Lords of the Admiralty, unless, of course, you are John Murray and can get away with it.

The physical and chemical measurements made during the voyage do not really belong here, but they led to an unexpected windfall that is hard to resist. Ocean temperature was measured during the voyage by specially designed thermometers that made readings at a series of depths at each station, giving the first temperature profiles of the ocean. The operation was rather delicate, as the ship had to be maneuvered almost continuously to keep the sounding rope vertical, so that the length run out was as close as possible to the real depth. These thermometer readings were summarized in the report, which consists mainly of voluminous data tables that list the thermometer readings over a depth gradient at each station. They showed that temperature decreased with depth, although not proportionately, as had been supposed, and that the deep sea is everywhere cold, just a little above freezing. These records remained the basis of knowledge about ocean temperatures for almost a century, as information accumulated from ships and buoys. In the early 1980s a new source of information became available when remote sensing from satellites was developed. This suggested that the surface of the oceans had warmed by an average of about 0.5 degrees Celsius during the twentieth century, but gave no information about the deep sea. This was eventually provided by the worldwide Argo program, which deployed automated floats drifting at specified depths to record temperature and then surface to beam the information to a satellite. It was then possible to compare the old *Challenger* data (1873–1876) with the new Argo measurements (2004–2010) to calculate the rise in temperature at all depths over the 135 years separating the two surveys. For the sea surface, the global average came out at 0.6 degrees Celsius, which is reassuringly close to estimates made by other methods, and also a rather striking testimony to the accuracy with which the crew of the *Challenger* were able to measure ocean temperature with thermometers strung on a rope from the heaving deck of a drifting ship. The warming decreases with depth, because the ocean is only slowly mixed by wind and current, becomes much less at 500 fathoms (about 0.1° C) and has disappeared at about 800 fathoms. It is reassuring to know that the deep sea has not yet been affected by global warming, tempered by the assurance that eventually it will be.

The New Zoology

Most of the reports describe the animals that were collected and cover the whole range of marine groups. They cover them very unevenly, though, because many of the groups that are most prominent in shallow water are much less abundant, or even completely absent, in the deep sea. If you explore the edge of the sea from a small boat, or with face mask and snorkel, you are likely to see familiar animals such as crabs, snapping shrimp, prawns, shoaling fish, fleshy sponges, large bivalves, sea slugs, classical short-armed starfish, stout-bodied, spiny sea urchins, chitons and squid, with a shark or two if you are lucky (or very unlucky). There will be reef corals in warmer seas, sea anemones and tunicates on the rocks, and bryozoans encrusting the blades of seaweed. None of these are common in the deep sea, where they are replaced by other groups of animals, or represented only by versions so greatly modified as to be almost unrecognizable.

To begin with, there are more echinoderms in the deep sea. Only a few percent of the large animals collected close to the shore will be starfish, sea urchins, or their relatives, but they become more prominent in deeper water, until they often make up one-third or more of the large animals brought up by the dredge from abyssal depths. There are different kinds of echinoderms, too, in deep water: the fantastically elaborate basket stars, hovering ghostlike holothurians such as *Peniagone*, thin-shelled, bottle-shaped echinoids such as *Pourtalesia*, and of course the emblematic stalked crinoids. Sponges are also common at all depths, but the soft or fibrous sponges of shallow seas are replaced by the fragile towers of glass sponges further down. Stony reef corals cannot live in deep water, but solitary cup corals are plentiful, and there may be large gardens of soft corals, the gorgonians, and pennatulids. Scaphopod mollusks are no longer a curiosity but a prominent member of the deepwater community. There are giant scavenging amphipods and strangely shaped isopods, unknown from the upper waters. There are large pycnogonids and a host of tiny cumacean crustaceans. There are more familiar animals such as cephalopods, but they are the peculiar cirroteuthid octopuses; there are fish, too, but they are mostly long-bodied, big-headed species with little resemblance to parrot fish or mullet. Above all, the sea floor is covered with the enigmatic xenos and komokis, the branching rhizopod tubes that were often draped across the mouth of the dredge and the mudballs inside it, which have no equivalent in shallow water and even today remain poorly understood. All of these were discovered, or first systematically collected, by the *Challenger*, and described in the reports. For the first time, the outline of the animal communities that inhabit half the surface of the Earth began to take shape.

The New Bodies

These unfamiliar animals often took unfamiliar forms. They might be unfamiliar only because they are antique, if they really are the last few survivors of groups that flourished long ago, and today linger on only in the calm, remote, unchanging world of the deep sea. Stalked crinoids first suggested this idea, and Willemoes-Suhm's blind lobster and the peculiar trilobite-like crustacean *Nebaliopsis* that they fished up from abyssal red clay in the middle of the Pacific seem to support it. So do the segmented monoplacophoran mollusks that the expedition failed to find, unless the drab limpet-like shell they found near Fiji was one. But there are just as many living fossils from the upper seas, such as the pearly nautilus and *Spirula*, which recall the coiled cephalopods of the Mesozoic. *Cephalodiscus*, which is now thought to be the lineal descendant of graptolites, lives in coastal waters. The bivalve *Trigonia* and the Port Jackson shark are found in harbors, and lungfish live in swamps. The characteristic animals of the abyss include glass sponges, which certainly evolved very early, but also include modern echinoderms such as brittle stars, which appear much later in the fossil record. The notion that the deep sea is the attic of the world lives on, perhaps because it has a certain dramatic appeal, but the evidence is ambiguous at best.

The alternative explanation is that deep-sea animals look unfamiliar because they are adapted to unfamiliar conditions of life. Living in the calm of the deep sea, they are freed from the need to have robust bodies and thick shells capable of withstanding wave and current. Animals like echinoids and bivalves that have stout tests or shells in shallow water evolve thin, fragile coverings to save the unnecessary expense of armor plate. Animals such as octopus and holothurians that are solid, muscular creatures in shallow seas evolve toward a translucent gelatinous body. Many deep-sea animals are blind, because eyes degenerate in the perpetual darkness where natural selection no longer acts to maintain vision. Others have evolved in the opposite direction because they use bioluminescent organs to signal to one another in the darkness, and have eyes of extraordinary acuity. They are in turn vulnerable to predators whose eyes can detect the faintest gleam of light and may become grotesquely modified, such as the photon-detection apparatus of the peculiar fish *Ipnops*. Blind animals such as the tripod fish *Bathypterois* may have long, sensitive filaments to detect motion in the place of eyes. The thick, fine ooze is apt to clog the apparatus of filter feeders such as sponges, bryozoans, or crinoids, so they evolve long stems or stalks to reach into clearer water. Many motile animals such as the deep-sea pycnogonids have long, slender legs to support themselves on the treacherous surface. All these are features in a distinctive suite of characters displayed by animals that live in the deep sea. They are not a reminiscence of ancient times, retained by archaic forms that have retired from the rough-and-tumble of life near the surface to the peace and quiet of the deeps. The stalked crinoids that started this train of thought have not retained their stalk as a memento of past times; they use it to prevent their food grooves from being clogged with mud. The notion that the deep sea is the lumber room of zoology exercised a strange fascination for a long time, but it was swept aside by the collections of the expedition. The features common to many deep-sea animals can be readily understood as specific adaptations to the environment they live in, rather than some mysterious remembrance of times past.

The New Way of Life

Some ways of life are not possible in the deep sea. Green plants cannot grow there, so there is no primary production, except in places (undetected by *Challenger*) where streams of minerals that can be metabolized by bacteria emerge on the seabed. The community as a whole must therefore be sustained by corpses that sink from the surface, either as a thin continuous rain of small particles or, as I have emphasized, as the rare descent of a large, dead animal. The sediment itself can be consumed faster than it accumulates, so that only its topmost skin contains much nutrition. It is harvested by minute crustaceans, the cumaceans and tanaids, able to scrape the thin bacterial film from the particles of ooze. Larger animals simply swallow the ooze wholesale, even though it is such thin broth. Deposit feeders such as echinoids or holothurians must feed and excrete continuously in order to move the sediment through their gut fast enough to maintain life. The sediment is bound by the xenos and komokis, which may be the primary consumers of the rain of dead plankton and the primary diet of deposit feeders, although their place in the economy of the deep sea is still uncertain.

Other ways of life evolve as novelties in the deep sea. Bacteria and small protozoans are much less abundant than they are at the surface, and filter feeding becomes correspondingly less profitable. Larger animals (although still minute) such as cumacean and tanaid crustaceans are more readily available, and filter feeders such as glass sponges, bivalves, and tunicates have, improbably, evolved to catch and consume them. Then there are ways of life that are followed almost everywhere but are particularly prominent in the deep sea. There are scavengers like vultures or hagfish in most places, for example, but the giant, blind amphipods that strip the carcasses of dead whales are much more important members of the community than their counterparts elsewhere.

The reports documented this new community of deep-sea animals, described their often bizarre structure, and opened the way to understanding the ecology of the abyssal plains. They continued to appear for 10 years and continued to be distributed free of charge to institutions around the world, which undoubtedly added to their influence. At one point, the Admiralty jibbed at the expense, but Murray forced them to reconsider by offering to pay out of his own pocket. It was the international cast of specialists, the timely preparation of authoritative reports, and the wide distribution of the annual volumes that raised the expedition from an interesting Victorian survey to the foundation of a new science of the oceans. None of this was inevitable. The specimens might have been distributed parochially to the staff of the British Museum, but for the vision of Wyville Thomson and the Royal Society; the reports might have been delayed for almost any length of time, but for the energy of John Murray; and the reports themselves might have been too expensive for any but a handful of foreign institutions, except for the generosity, however grudging, of the Treasury. Without all these, the voyage would have remained, with a dozen others, as a footnote in history. I doubt that this would have made much difference in the long run because there were several other expeditions planned or under way with similar objectives at the time. But it would have taken longer, perhaps much longer, without the vision and energy of the men who planned the expedition and brought it to a successful conclusion, besides the dedication of the rather amateurish cast of naturalists who, week after week for three and a half years, and in all weathers, picked through the piles of mud brought up by the dredge.

The British Museum finally got its specimens, and duplicates were sent to dozens of other museums. Some specimens, especially from shallow trawls, had already been donated to local museums during the course of the voyage. Others were distributed rather haphazardly from the Challenger Office, and the authors of the reports themselves sometimes failed to return all the material that had been sent to them. The British Museum itself eventually redistributed some thousands of specimens from the *Challenger* collections it had so earnestly claimed. It might be worth checking to see if your own local museum received anything. I was recently surprised to find *Challenger* material a few feet from my own office in the Redpath Museum, at McGill University in Montreal; John Murray had sent some manganese nodules from the abyssal plain of the Pacific as a gift to Sir William Dawson, who was the Director at the time. The *Challenger* may be closer than you think.

References

You might want to check the sources that I have used in writing this book, or to read more about the expedition or the animals. I shall first list the contemporary sources, written by people who were actually on the ship. I will then list the books and reviews that have subsequently appeared. All of these can be read for pleasure. The bulk of the references follow, organized by section, so that you can check what I have written against the primary literature.

Before I begin, though, the indispensable sources are the two books by Murray.

Wyville Thomson, C. and Murray, J. 1885. *Report on the Scientific Results of the Voyage of HMS Challenger during the years 1873–76: Narrative.* 2 vols. (in 3 parts). Her Majesty's Stationery Office, London.

Wyville Thomson, C. and Murray, J. 1885. *Report on the Scientific Results of the Voyage of HMS Challenger during the years 1872–76: A Summary of the Scientific Results.* 2 vols. (in 3 parts). Her Majesty's Stationery Office, London.

The reports themselves can be read at

http://www.19thcenturyscience.org/HMSC/HMSC-INDEX/index-linked.htm.

You might also want to consult the reports of the Danish *Galathea* expedition:

Galathea Reports: Scientific Results of the Danish Deep-Sea Expedition round the World, 1950–1952. Galathea Committee, Copenhagen, 1956–1959

There are also several contemporary accounts of the voyage.

Campbell, Lord George. 1876. *Log Letters from "The Challenger".* Macmillan, London.

Campbell, Lord George. No date (?1875). Private Journal of Lord George Granville Campbell, H.M.S. "Challenger," from the Cape of Good Hope to Australia. Printed for the use of his family and friends only.

Moseley, H. N. 1879. *Notes by a Naturalist: An Account of Observations made during the Voyage of H.M.S. "Challenger" round the World in the years 1872–1876.* John Murray, London.

Spry, W. J. J. 1877. *The Cruise of Her Majesty's Ship "Challenger": Voyages over many Seas, Scenes in many Lands.* Sampson Low, Marston, Searle & Divington, London.

Swire, H. 1938. *The Voyage of the* Challenger: *A Personal Narrative of the Historic Circumnavigation of the Globe in the Years 1872–1876.* Golden Cockerel Press, London.

Wild, J. J. 1878. *At Anchor: A Narrative of Experiences Afloat and Ashore during the Voyage of H.M.S.* Challenger, *from 1872 to 1876.* M. Ward, London.

Wyville Thomson, C. 1877. *The Atlantic: A Preliminary Account of the general results of the exploring voyage of H.M.S. "Challenger" during the year 1873 and the early part of the year 1876.* 2 vols. Macmillan, London.

Matkin's letters home have been published:

Rehbock, P. F. (editor). 1992. *At Sea with the Scientifics: The* Challenger *Letters of Joseph Matkin.* University of Hawaii Press, Honolulu.

I have also consulted these unpublished journals:

Journal of Assistant Paymaster John Hynes Dec. 1873–Mar. 1874. Caird Library JOD/15/1.
Journal and Remark Book of A. F. Balfour. Foyle Library of Royal Geographical Society ABA/3.
Journal of Paymaster R. R. A. Richards. Foyle Library of Royal Geographical Society.

There are also several books written after the voyage by nonparticipants.

Linklater, E. 1972. *The Voyage of the* Challenger. John Murray, London.
MacDougall, D. 2019. *Endless Novelties of Extraordinary Interest: The Voyage of H.M.S.* Challenger *and the Birth of Modern Oceanography.* Yale University Press, New Haven, CT.
Aitken, F. and Foulc, J.-N. 2019. *From Deep Sea to Laboratory: The First Exploration of the Deep Sea by H.M.S.* Challenger *(1872–1876).* ISTE and John Wiley, London and Hoboken, NJ.

Some of the illustrations from the reports have been published separately.

Haeckel, E. No date (from Reports of 1882, 1887, 1888, 1889). *Art Forms from the Abyss.* Prestel, Munich.
Haeckel, E. No date (from Reports of 1882, 1887, 1888, 1889). *Art Forms from the Ocean.* Prestel, Munich.

There are two recent books that make extensive references to the *Challenger* expedition.

Corfield, R. 2003. *The Silent Landscape: The Scientific Voyage of HMS* Challenger. Joseph Henry Press, Washington, DC.
Rozwadowski, H. M. 2005. *Fathoming the Ocean: The Discovery and Exploration of the Deep Sea.* Belknap Press, Harvard University Press, Cambridge, MA.

There are many reviews of the voyage. Here are a few.

Brunton, E. V. 1994. *The* Challenger *Expedition, 1872–1876: A Visual Index.* The Natural History Museum, London: Historical Studies in the Life and Earth Sciences No. 2.
Charnock, H. 1972. HMS *Challenger* and the development of marine science: The Duke of Edinburgh Lecture. *Journal of Navigation* 26: 1–12.
Codling, R., 1997. HMS *Challenger* in the Antarctic: Pictures and photographs from 1874. *Landscape Research* 22(2): 191–208.
Goode, G. Brown. 1884. The exploring voyage of the *Challenger. Science* 4: 176–179.
Hedgpeth, J. W. 1946. The voyage of the *Challenger. Scientific Monthly* 63: 194–202.
Yonge, M. 1971. The inception and significance of the *Challenger* expedition. *Proceedings of the Royal Society of Edinburgh* 72: 1–13.

There are also several useful websites, especially the HMS Challenger Project of the Royal Albert Memorial Museum, Exeter, UK:
https://rammhmschallenger.wordpress.com/about-us/.

Articles in the primary literature

These are articles, most of them in the primary literature, that give additional information about animals described in the text. For background information I habitually use Libby

Hyman's famous text (Hyman, L. H. 1940–1967. *The Invertebrata*. 5 vols. McGraw-Hill, New York). It is out of date and incomplete (no arthropods, for example) because she retired before it could be finished, but it remains my first port of call.

Chapter 1

Two monographs on oceanography were published by Wyville Thomson and Murray, the one before and the other after the expedition:

Murray, J. and Hjort, J. 1912. *The Depths of the Ocean*. Macmillan, London.
Wyville Thomson, C. 1873. *The Depths of the Sea*. Macmillan, London.

Here are some more recent monographs:

Gage, J. D. and Tyler, P. A. 1991. *Deep-Sea Biology: A Natural History of the Organisms at the Deep-Sea Floor*. Cambridge University Press, Cambridge.
Herring, P. 2002. *The Biology of the Deep Ocean*. Oxford University Press, Oxford.
Rex, M. A. and Etter, R. J. 2010. *Deep-Sea Biodiversity: Pattern and Scale*. Harvard University Press, Cambridge, MA.
Rowe, G. T. (editor). 2005. *Deep-Sea Biology*. Vol. 8. Harvard University Press, Cambridge, MA.
Tyler, P. A. (editor). 2003. *Ecosystems of the Deep Oceans*. Elsevier, Amsterdam.

There are also some useful or provocative short reviews.

Danovaro, R., Snelgrove, P. V. R., and Tyler, P. 2014. Challenging the paradigms of deep-sea ecology. *Trends in Ecology and Evolution* 29: 465–475.
Ramirez-Llodra, E. et al. 2010. Deep, diverse and definitely different: Unique attributes of the world's largest ecosystem. *Biogeosciences* 7: 2851–2899.

Chapter 2

Anderson, T. R. and Rice, T. 2006. Deserts on the sea floor: Edward Forbes and his azoic hypothesis for a lifeless deep ocean. *Endeavour* 30: 131–137.
Forbes, E. 1835. Records of the results of dredging. *Magazine of Natural History* 8: 68–69.
Forbes, E. 1844. On the light thrown on geology by submarine researches. *Edinburgh New Philosophical Journal* 36: 318–326.
Forbes, E. 1844. Report on the Mollusca and Radiata of the Aegean Sea, and on their distribution, considered as bearing on geology. Report of the Thirteenth Meeting of the British Association for the Advancement of Science held at Cork in August 1843, 130–193. John Murray, London.
Merriman, D. 1965. Edward Forbes—Manxman. *Progress in Oceanography* 3: 191–206.
Rehbock, P. F. 1979. The early dredgers: "Naturalizing" in British seas, 1830–1850. *Journal of the History of Biology* 12: 293–368.

Chapter 3

Burstyn, H. L. 1971. Pioneering in large-scale scientific organization: The *Challenger* expedition and its report. I. Launching the expedition. *Proceedings of the Royal Society of Edinburgh* 72: 47–61.
Carpenter, W. B. and Gwyn Jeffreys, J. 1870. Report on deep-sea researches carried on during the months of July, August and September 1870, in H.M. Surveying Ship "Porcupine". *Proceedings of the Royal Society of London* 19: 145–221.
Carpenter, W. B. and Wyville Thomson, C. 1868. Preliminary report of dredging operations in the seas to the north of the British Islands, carried on in Her Majesty's Steam-Vessel *Lightning*. *Proceedings of the Royal Society of London* 17: 168–200.

Report of the Circumnavigation Committee of the Royal Society. 1872. 15 pages. Government Printing Office, Washington, DC. The original is from the Minutes of the meeting of the Council of the Royal Society, 14 November 1872.

Chapter 4

The original narrative of the voyage gives the names and ranks of all the officers on board.

Rice, A. L. 1989. Oceanographic fame—and fortune: The salaries of the sailors and scientists on HMS *Challenger*. *Archives of Natural History* 16: 213–220.

Chapter 5

A Bright Red Starfish (*Hymenaster*)

Gale, K. S. P., Hamel, J.-F., and Mercier, A. 2013. Trophic ecology of deep-sea Asteroidea (Echinodermata) from eastern Canada. *Deep-Sea Research I* 80: 25–36.

Mah, C. 2017. Defenses in sea stars. Ocean Exploration and Research. NOAA website at https://oceanexplorer.noaa.gov/okeanos/explorations/ex1703/logs/mar24/welcome.html.

Gorgonians (*Strophogorgia*)

Sanchez, J. A. 2004. Evolution and dynamics of branching colonial form in modular cnidarians: Gorgonian octocorals. *Hydrobiologia* 530: 283–290.

Pennatulids (*Umbellula*, Pennatulacea)

Paterson, G. L. J., Glover, A. G., Barrio Frojan, C. R. S., Whitaker, A., Budaeva, N., Chimonides, J., and Doner, S. 2009. A census of abyssal polychaetes. *Deep-Sea Research II* 56: 1739–1746.

Tyler, P. A., Bronsdon, S. K., Young, C. M., and Rice, A. L. 1995. Ecology and gametogenic biology of the genus Umbellula (Pennulatacea) in the North Atlantic Ocean. *Internationale Revue der gesamten Hydrobiologie und Hydrographie* 80: 187–199.

Chapter 6

Globigerina

Dulkiewicz, A., Muller, R. D., O'Callaghan, S., and Jonasson, H. 2015. Census of seafloor sediments in the world's ocean. *Geology* 43: 795–798.

Red Clay

Boudreau, B. P., Middelburg, J. J., and Meysman, F. J. R. 2010. Carbonate compensation dynamics. *Geophysical Research Letters* 37: L03603.

The Mid-Atlantic Ridge

Ballard, R. D. and Moore, J. G. 1977. *Photographic Atlas of the Mid-Atlantic Rift Valley*. Springer-Verlag, New York.

Smith, D. K. and Cann, J. R., 1993. Building the crust at the Mid-Atlantic Ridge. *Nature* 365: 707–715.

A Blind Lobster (*Willemoesia*, Decapoda)

Bezerra, L. E. A. and Ribeiro, F. B. 2015. Primitive decapods from the deep sea: First record of blind lobsters (Crustacea: Decapoda: Polychelidae) in northeastern Brazil. *Nauplius* 23: 125–131.

Black Pebbles

Baturin, G. N. 1988. *The Geochemistry of Manganese and Manganese Nodules in the Ocean.* D. Reidel, Dordrecht.

Bender, M. L., Ku, T. L., and Broecker, W. S. 1966. Manganese nodules: Their evolution. *Science* 151: 325–328.

A Deep Worm (Myriochele, Polychaeta)

Kupriyanova, E. K. et al. 2014. Serpulids living deep: Calcareous tubeworms beyond the abyss. *Deep-Sea Research I* 90: 91–104.

Paterson, G. L. J. et al. 2009. A census of abyssal polychaetes. *Deep-Sea Research II* 56: 1739–1746.

Snails with Wings (pteropods, Gastropoda)

Burridge, A. K., Goetze, E., Wall-Palmer, D., Le Double, S. L., Huisman, J., and Peijnenburg, K. T. C. A. 2017. Diversity and abundance of pteropods and heteropods along a latitudinal gradient across the Atlantic Ocean. *Progress in Oceanography* 158: 213–223.

Kubilius, R. A., Kohnert, P., Brenzinger, B., and Schrodl, M. 2014. 3d-microanatomy of the straight-shelled pteropod *Creseis clava* (Gastropoda: Heterobranchia: Euthecosomata). *Journal of Molluscan Studies* 80: 585–603.

Melkert, M. J., Ganssen, G., Helder, W., and Troelstra, S. R. 1992. Episodic preservation of pteropod oozes in the deep Northeast Atlantic Ocean: Climatic change and hydrothermal activity. *Marine Geology* 103: 407–422.

A Stalked Crinoid (*Rhizocrinus*, Crinoidea)

Ameziane, N. and Roux, M. 1997. Biodiversity and historical biogeography of stalked crinoids (Echinodermata) in the deep sea. *Biodiversity and Conservation* 6: 1557–1570.

Macurda, D. B. and Meyer, D. L. 1976. The identification and interpretation of stalked crinoids (Echinodermata) from deep-water photographs. *Bulletin of Marine Science* 26: 205–215.

Macurda, D. B. and Meyer, D. L. 1976. The morphology and life habits of the abyssal crinoid Bathycrinus aldrichianus Wyville Thomson and its paleontological implications. *Journal of Palaeontology* 50: 647–667.

The Attic of the World?

Laflamme, M. and Narbonne, G. M. 2008. Ediacaran fronds. *Palaeogeography, palaeoclimatology, palaeoecology* 258: 162–179.

Menzies, R. J. and Imbrie, J. 1958. On the antiquity of the deep sea bottom fauna. *Oikos* 9: 192–210.

Thuy, B., Gale, A. S., Kroh, A., Kucera, M., Numberger-Thuy, L. D., Reich, M., and Stohr, S. 2012. The ancient origin of the modern deep-sea fauna. *PLoS One* 7: e46913.

Wilson, G. D. F. 1999. Some of the deep-sea fauna is ancient. *Crustaceana* 72: 1019–1030.

Zenkevitch, L. A. and Birstein, J. A. 1960. On the problem of the antiquity of the deep-sea fauna. *Deep-Sea Research* 7: 10–23.

Chapter 7

Sargassum

Rowe, G. T. and Staresinic, N. 1979. Sources of organic matter to the deep-sea benthos. *Ambio Special Report* 6: 19–23.

Ryland, J. S. 1974. Observations on some epibionts of gulf-weed, *Sargassum natans* (L.) Meyen. *Journal of Experimental Biology and Ecology* 14: 17–25.

Weis, J. S. 1968. Fauna associated with pelagic *Sargassum* in the Gulf Stream. *American Midland Naturalist* 80: 554–558.

Sea Serpents
For an introduction to sipunculids, consult a text such as Hyman. The Wikipedia entry is sound.

The Crinoid Plague
For an introduction to myzostomids, consult a text such as Hyman. The Wikipedia entry is brief and incomplete but sound.
Helm, C., Weigert, A., Mayer, G., and Bleidom, C. 2013. Myoanatomy of *Myzostoma cirriferum* (Annelida, Myzostomida): Implications for the evolution of the myzostomid body plan. *Journal of Morphology* 274: 456–466.

Brittle Stars
Metaxas, A. and Giffin, B. 2004. Dense beds of the ophiuroid *Ophiacantha abyssicola* on the continental slope off Nova Scotia, Canada. *Deep-Sea Research I* 51: 1307–1317.

Sea Legs (Pycnogonida, Arthropoda)
Arnaud, F. and Bamber, R. N. 1987. The biology of Pycnogonida. *Advances in Marine Biology* 24: 2–96.
Mercier, A., Baillon, S., and Hamel, J.-F. 2015. Life history and feeding biology of the deep-sea pycnogonid *Nymphon hirtipes*. *Deep Sea Research I* 106: 1–8.
Wicksten, M. K. 2017. Feeding on cnidarians by giant pycnogonids (Pycnogonida: Colossendeidae Jarzinsky 1870) in the North Central Pacific and North Atlantic Oceans. *Journal of Crustacean Biology* 37: 359–360.

A Worm in the Wrong Place (*Geonemertes*, Nemertea)
Jones, H. D. and Sterrer, W. 2005. Terrestrial planarians (Platyhelminthes, with three new species) and nemertines of Bermuda. *Zootaxa* 1001: 31–58.
Moore, J. and Gibson, R. 1985. The evolution and comparative physiology of terrestrial and freshwater nemerteans. *Biological Reviews* 60: 257–312.

Chapter 8
Scatter (bristlemouths, Osteichthyes)
D'Elia, M., Warren, J. D., Rodriguez-Pinto, I., Sutton, T. T., and Boswell, K. M. 2016. Diel variation in the vertical distribution of deep-water scattering layers in the Gulf of Mexico. *Deep Sea Research I* 115: 91–102.
Dietz, R. S. 1962. The sea's deep scattering layers. *Scientific American* 207: 44–51.
Judkins, D. C. and Haedrich, R. L. 2018. The deep scattering layer micronektonic fish faunas of the Atlantic mesopelagic ecoregions with comparison of the corresponding decapod shrimp faunas. *Deep-Sea Research I* 136: 1–30.

Blink (bioluminescence)
Anctil, M. 2018. *Luminous Creatures*. McGill-Queen's University Press, Montreal.
Haddock, S. H. D., Moline, M. A., and Case, J. F. 2010. Bioluminescence in the sea. *Annual Reviews of Marine Science* 2: 443–493.
Land, M. F. 2000. On the functions of double eyes in midwater animals. *Philosophical Transactions of the Royal Society of London B* 355: 1147–1150.

Dwarf Males (*Idiacanthus*, Osteichthyes; *Scalpellum*, Cirripedia)

Clarke, T. A. 1983. Sex ratios and sexual differences in size among mesopelagic fishes from the central Pacific Ocean. *Marine Biology* 73: 203–209.

Dreyer, N., Hoeg, J. T., Hess, M., Sorensen, S., Spremberg, U., and Yusa, Y. 2018. When dwarf males and hermaphrodites copulate: First record of mating behaviour in a dwarf male using the androdioecious barnacle *Scalpellum scalpellum* (Crustacea: Cirripedia: Thoracica). *Organisms, Diversity and Evolution* 18: 115–123.

Spremberg, U., Hoeg, J. T., Buhl-Mortensen, L., and Yusa, Y. 2012. Cypris settlement and dwarf male formation in the barnacle *Scalpellum scalpellum*: A model for an androdioecious reproductive system. *Journal of Experimental Marine Biology and Ecology* 422: 39–47.

Volrath, F. 1998. Dwarf males. *Trends in Ecology and Evolution* 13: 159–163.

Weedfall

Fleury, J. G. and Drazen, J. C. 2013. Abyssal scavenging communities attracted to *Sargassum* and fish in the Sargasso Sea. *Deep-Sea Research I* 72: 141–147.

Slaters (Isopoda, Pancrustacea)

Hessler, R. R., Wilson, G. D., and Thistle, D. 1979. The deep-sea isopods: A biogeographic and phylogenetic overview. *Sarsia* 64: 67–75.

A Frangible Bag (*Pyrosoma*, Tunicata)

Griffin, D. J. G. and Yaldwyn, J. C. 1970. Giant colonies of pelagic tunicates (*Pyrosoma spinosum*) from SE Australia and New Zealand. *Nature* 226: 464–465.

Mackie, G. O. and Bone, Q. 1978. Luminescence and associated effector activity in *Pyrosoma* (Tunicata, Pyrosomida). *Proceedings of the Royal Society B* 202: 483–495.

Van Soest, R. W. M. 1981. A monograph of the order Pyrosomatida (Tunicata, Thaliacea). *Journal of Plankton Research* 3: 603–631.

Haeckel's Mistake (Xenophyophorea)

Gooday, A. J. and Jorissen, F. J. 2012. Benthic foraminiferal biogeography: Controls on global distribution patterns in deep-water settings. *Annual Reviews of Marine Science* 4: 237–262.

Gooday, A. J., Holzmann, M., Caulle, C., Goineau, A., Kamenskaya, O., Weber, A. A.-T., and Pawlowski, J. 2017. Giant protists (xenophyophores, Foraminifera) are exceptionally diverse in parts of the abyssal Pacific Ocean licensed for polymetallic nodule exploration. *Biological Conservation* 207: 106–116.

Kamenskaya, O. E., Gooday, A. J., Tendal, O. S., and Melnik, V. F. 2015. Xenophyophores (Protista, Foraminifera) from the Clarion-Clipperton Fracture Zone, with descriptions of three new species. *Marine Biodiversity* 45: 581–593.

Levin, L. S. 1994. Paleoecology and ecology of xenophyophores. *Palaios* 9: 32–41.

Absence of a Gutless Worm (Pogonophora)

Southward, E. C., Schulze, A., and Gardiner, S. L. 2005. Pogonophora (Annelida): Form and function. Pp. 227–251 in Bartolomaeus, T. and Purschke, E. (editors), *Morphology, Molecules, Evolution and Phylogeny in Polychaeta and Related Taxa*. Springer, Dordrecht.

A Cup of Coral

Cairns, S. D. 2007. Deep-water corals: An overview with special reference to diversity and distribution of deep-water scleractinian corals. *Bulletin of Marine Science* 81: 311–322.

Chapter 9

Reynolds, P. D. 2002. The Scaphopoda. *Advances in Marine Biology* 42: 137–236.

The Sea Devil (*Ceratias*, Osteichthyes)
Munk, O. 1999. The escal photophore of ceratioids (Pisces; Ceratioidei): A review of structure and function. *Acta Zoologica* 80: 265–284.

A Fierce Clam (*Rhinoclama*, Bivalvia)
Allen, J. A. and Morgan, R. E. 1981. The functional morphology of Atlantic deep water species of the families Cuspidariidae and Poromyidae (Bivalvia): An analysis of the evolution of the septibranch condition. *Philosophical Transactions of the Royal Society B* 294: 413–546.

Gennadas (Macrura, Decapoda)
Vereshchaka, A. L., Lunina, A. A., and Olesen, J. 2017. The genus *Gennadas* (Benthesicymidae: Decapoda): Morphology of copulatory characters, phylogeny and coevolution of genital structures. *Royal Society Open Science* 4: 171288.

A Cloak of Invisibility (*Cystisoma*, Amphipoda)
Bagge, L. E., Osborn, K. J., and Johnsen, S. 2016. Nanostructures and monolayers of spheres reduce surface reflections in hyperiid amphipods. *Current Biology* 26: 3071–3076.
Brusca, G. J. 1981. On the anatomy of *Cystisoma* (Amphipoda: Hyperiidea). *Journal of Crustacean Biology* 1: 358–375.
Cronin, T. W. 2016. Camouflage: Being invisible in the open ocean. *Current Biology* 26: R1179–R1180.
Mills, E. L. 1971. T.R.R. Stebbing, the Challenger and knowledge of deep-sea Amphipoda. *Proceedings of the Royal Society of Edinburgh* 72: 68–87.

Gill Slits (*Glandiceps*, Hemichordata)
Holland, N. D., Osborn, K. J., Gebruk, A. V., and Rogacheva, A. 2013. Rediscovery and augmented description of the HMS "*Challenger*" acorn worm (Hemichordata, Enteropneusta), *Glandiceps abyssicola*, in the equatorial Atlantic abyss. *Journal of the Marine Biological Association of the U.K.* 93: 2197–2205.

Peniagone (Holothuria, Echinodermata)
Cross, I. A., Gebruk, A., Billett, D., and Rogacheva, A. 2012. Rediscovery of the elpidiid holothurian *Peniagone horrifer* (Elaspodida, Holothuroidea: Echinodermata) in the southern Indian Ocean. *Marine Biodiversity* 42: 241–245.
Smith, K. L., Kaufmann, R. S., and Wakefield, W. W. 1991. Mobile megafaunal activity monitored with a time-lapse camera in the abyssal North Pacific. *Deep-Sea Research I* 40: 2307–2324.

Moss Animals and Lamp Shells (Bryozoa and Brachiopoda)
Hayward, P. J. 1981. The Cheilostomata (Bryozoa) of the deep sea. *Galathea Reports* 15: 21–68.
Okamura, B. and Doolan, L. A. 1933. Patterns of suspension feeding in the freshwater bryozoan *Plumatella repens*. *Biological Bulletin* 184: 52–56.

Naming

The rules for naming animals are set out on the website of the International Commission on Zoological Nomenclature:
https://www.iczn.org/the-code/the-international-code-of-zoological-nomenclature/the-code-online/.

Sally Lightfoot (*Grapsus*, Decapoda)

Freire, A. S., Pinheiro, M. A. A., Karam-Silva, H., and Teschima, M. M. 2011. Biology of *Grapsus grapsus* (Linnaeus, 1758) (Brachyura, Grapsidae) in the Saint Peter and Saint Paul Archipelago, equatorial Atlantic Ocean. *Helgoland Marine Research* 65: 263–273.

Tripod (*Bathypterois*, Osteichthyes)

Sulak, K. J. 1977. *The systematics and biology of* Bathypterois (Pisces, Chlorophthalmidae). *Galathea Reports* 14: 49–108.

Fine Filtration (*Geodia*, Demospongia)

Burns, E., Ifrach, I., Carmell, S., Pawlik, J. R., and Ilan, M. 2003. Comparison of anti-predatory defenses of Red Sea and Caribbean sponges: I. Chemical defense. *Marine Ecology Progress Series* 252: 105–114.

Burns, E. and Ilan, M. 2003. Comparison of anti-predatory defenses of Red Sea and Caribbean sponges: II. Physical defense. *Marine Ecology Progress Series* 252: 115–123.

Leys, S. P., Yahel, G., Reidenbach, M. A., Tunnicliffe, V., Shavit, U., and Reiswig, H. M. 2011. The sponge pump: The role of current induced flow in the design of the sponge body plan. *PLoS One* 6: e27787.

Reiswig, H. M. 1975. Bacteria as food for temperate-water marine sponges. *Canadian Journal of Zoology* 53: 582–589.

Yin, Z., Zhu, M., Davidson, E. H., Bottjer, D. J., Zhao, F., and Tafforeau, P. 2015. Sponge grade body fossil with cellular resolution dating 60 Myr before the Cambrian. *Proceedings of the National Academy of Science of the U.S.A.* 112: E1453–E1460.

Chapter 10

Sea skaters (*Halobates*, Insecta)

Andersen, N. M. and Cheng, L. 2002. The marine insect *Halobates* (Heteroptera: Gerridae): Biology, adaptations, distribution, and phylogeny. *Oceanography and Marine Biology Annual Review* 42: 119–180.

Cheng, L. 1985. Biology of *Halobates* (Heteroptera, Gerridae). *Annual Reviews of Entomomology* 30: 111–135.

A Cosmopolitan Cucumber (*Psychropotes*, Holothuroidea)

Gubili, C. et al. 2017. Species diversity in the cryptic abyssal holothurian *Psychropotes longicauda* (Echinodermata). *Deep-Sea Research II* 137: 288–296.

Neither Blind nor Sighted (*Ipnops*, Osteichthyes)

Munk, O. 1959. The eyes of *Ipnops murrayi* Gunther, 1878. *Galathea Reports* 3: 79–87.

A Tale of Two Brothers

Rosenthal, E. 1952. *Shelter from the Spray: Being the True and Surprising story of Frederick and Gustav Stoltenhoff, Lately of Cape Town, Their Various Adventures on a Desert Island and Elsewhere, Their Love Affairs and Subsequent Fate.* Howard Timmins, Cape Town.

Snorkel Starfish (*Porcellanaster*, Asteroidea)

Mironov, A. N., Dilman, A. B., Vladychenskaya, I. P., and Petrov, N. B. 2016. Adaptive strategy of the porcellanasterid sea stars. *Biology Bulletin* 43: 503–516.

Chapter 11

Sea Serpent (ocean sunfish, *Mola*, Osteichthyes)

Pope, E. C. et al. 2010. The biology and ecology of the ocean sunfish *Mola mola*: A review of current knowledge and future research perspectives. *Reviews in Fish Biology and Fisheries* 20: 471–487.

Cucumbers with Legs (*Ellipinion*, Holothuroidea)

Gebruk, A. 1996. Locomotory organs in the elasipodid holothuriuans: Functional and evolutionary approaches. *Echinoderm Research* 1995: 95–102.

O'Loughlin, P. M., Paulay, G., Davey, N., and Michonneau, F. 2011. The Antarctic region as a marine biodiversity hotspot for echinoderms: Diversity and diversification of sea cucumbers. *Deep-Sea Research II* 58: 264–275.

Grenadiers and Cutthroats (*Coryphaenoides* and *Synaphobranchus*)

Gaither, M. R. et al. 2016. Depth as a driver of evolution in the deep sea: Insights from grenadiers (Gadiformes: Macrouridae) of the genus *Coryphaenoides*. *Molecular Phylogenetics and Evolution* 104: 73–82.

High Pressure

Blaxter, J. H. S. 1980. The effect of hydrostatic pressure on fishes. Pp. 369–386 in Ali, M. A. (editor), *Environmental Physiology of Fishes*. Springer, New York.

Children, J. J. and Somero, G. N. 1979. Depth-related enzymic activities in muscle, brain and heart of deep-living pelagic marine teleosts. *Marine Biology* 52: 273–283.

Drazen, J. C. and Haedrich, R. L. 2012. A continuum of life histories in deep-sea demersal fishes. *Deep-Sea Research I* 61: 34–42.

Gerringer, M. E., Drazen, J. C., Linley, T. D., Summers, A. P., Jamieson, A. J., and Yancey, P. H. 2017. Distribution, composition and functions of gelatinous tissues in deep-sea fishes. *Royal Society Open Science* 4: 171063.

Morita, T. 2010. High-pressure adaptation of muscle proteins from deep-sea fishes, Coryphaenoides yaquinae and C. armatus. *Annals of the New York Academy of Sciences* 10: 1111.

Neat, F. C. and Campbell, N. 2013. Proliferation of elongate fishes in the deep sea. *Journal of Fish Biology* 83: 1576–1591.

Wardle, C. S., Tettem-Lartey, N., MacDonald, A. G., Harper, A. A., and Pennec, J.-P. 1987. The effect of pressure on the lateral swimming muscle of the European eel Anguilla Anguilla and the deep-sea eel Histiobranchus bathybius: Results of Challenger cruise 6B/85. *Comparative Biochemistry and Physiology* 88A: 595–598.

Colossal (*Cirrothauma* and *Bathyteuthis*, Cephalopoda)

Collins, M. A. and Rodhouse, P. G. K. 2006. Southern Ocean cephalopods. *Advances in Marine Biology* 50: 191–264.

Rodhouse, P. G. and White, M. G. 1995. Cephalopods occupy the ecological niche of epipelagic fish in the Antarctic Polar Frontal Zone. *Biological Bulletin* 189: 77–80.

Rosa, R., Lopes, V. M., Guerreiro, M., Bolstad, K., and Xavier, J. C. 2017. Biology and ecology of the world's largest invertebrate, the colossal squid (*Mesonychoteuthis hamiltoni*): A short review. *Polar Biology* 40: 1871–1883.

The Peculiar Animals (echinoderms)
Hennebert, E., Haesaerts, D., Dubois, P., and Flammang, P. 2010. Evaluation of the different forces brought into play during tube foot activities in sea stars. *Journal of Experimental Biology* 213: 1162–1174.

Cabbage (*Pringlea*, Angiospermia)
Dorne, A. J. and Bligny, R. 1993. Physiological adaptation to subantarctic climate by the Kerguelen cabbage, *Pringlea antiscorbutica* R. Br. *Polar Biology* 13: 55–60.
Hatt, H. H. 1949. Vitamin C content of an old antiscorbutic: The Kerguelen cabbage. *Nature* 164: 1081–1082.

Transit of Venus
Perry, S. J. 1876. *Notes of a Voyage to Kerguelen Island to observe the transit of Venus, December 8, 1874.* H.S. King, London.

Cold Water
Iudicone, D., Speich, S., Madec, G., and Blanke, B. 2008. The global conveyor belt from a Southern Ocean perspective. *Journal of Physical Oceanography* 38: 1401–1425.

Warm water
Falkowski, P. G. et al. 2011. Ocean deoxygenation: Past, present and future. *Eos* 92: 409–420.
Smith, A. B. and Stockley, B. 2005. The geological history of deep-sea colonization by echinoids: Roles of surface productivity and deep-water ventilation. *Proceedings of the Royal Society B* 272: 865–869.

Salt Water
Epstein, F. H. 1999. The sea within us. *Journal of Experimental Biology* 284: 50–54.
Macallum, A. B., 1926. The paleochemistry of the body fluids and tissues. *Physiological Reviews* 6(2): 316–357.
Robertson, J. D., 1949. Ionic regulation in some marine invertebrates. *Journal of Experimental Biology* 26(2): 182–200.
Ruppert, E. E. and Carle, K. J. 1983. Morphology of metazoan circulatory systems. *Zopomorphology* 103: 193–208.

Bathydraco
Wohrman, A. P. A. 1998. Aspects of eco-physiological adaptations in Antarctic fish. Pp. 119–128 in di Prisco, G., Pisano, E., and Clarke, A. (editors), *Fishes of Antarctica*. Springer-Verlag, Rome.

Ice
Codling, R. 1997. HMS Challenger in the Antarctic: Pictures and photographs from 1874. *Landscape Research* 22: 191–208.

Diatom Ooze

Laney, S. R., Olson, R. J., and Sosik, H. M. 2012. Diatoms favor their younger daughters. *Limnology and Oceanography* 57: 1572–1578.

Spun Glass

Janussen, D., Tabachnick, K. R., and Tendal, O. S. 2004. Deep-sea Hexactinellida (Porifera) of the Weddell Sea. *Deep-Sea Research II* 51: 1857–1882.

Leys, S. P., Mackie, G. O., and Reiswig, H. M. 2007. The biology of glass sponges. *Advances in Marine Biology* 52: 2–145.

A Fierce Sponge (*Cladorhiza*, Demospongia)

Hestetun, J. T., Rapp, H. T., and Xavier, J. 2017. Carnivorous sponges (Porifera, Cladorhizidae) from the Southwest Indian Ocean Ridge seamounts. *Deep-Sea Research II* 137: 166–189.

Vaclet, J. 2007. Diversity and evolution of deep-sea carnivorous sponges. Pp. 107–115 in Custodio, M. R., Lobo-Hajdu, G., Hajdu, E., and Muricy, G. (editors), *Porifera Research: Biodiversity, Innovation and Sustainability.* Museu Nacional, Rio de Janeiro, Brazil.

Vaclet, J. and Boury-Esnault, N. 1995. Carnivorous sponges. *Nature* 373: 333–335.

False Witnesses (*Serolis*, Isopoda; monoplacophorans)

Schwabe, E. 2008. A summary of reports of abyssal and hadal Monoplacophora and Polyplacophora (Mollusca). *Zootaxa* 1866: 205–222.

Phronima

Laval, P.1980. Hyperiid amphipods as crustacean parasitoids associated with gelatinous zooplankton. *Oceanography and Marine Biology Annual Review* 18: 11–56.

The Enigmatic Tunic (*Abyssascidia*, Tunicata)

Nakashima, K., Yamada, L., Saton, Y., Azuma, J., and Satoh, N. 2004. The evolutionary origin of animal cellulose synthase. *Development Genes and Evolution* 214: 81–88.

Chapter 12

A Precious Clam (*Trigonia*, Bivalvia)

Gould, S. J. 1968. *Trigonia* and the origin of species. *Journal of the History of Biology* 1: 41–56.

Stanley, S. M. 1977. Coadaptation in the Trigoniidae, a remarkable family of burrowing bivalves. *Palaeontology* 20: 869–899.

A Shark with Molars (*Heterodontus*, Chondrichthyes)

Powter, D. M. and Gladstone, W. 2008. The reproductive biology and ecology of the Port Jackson shark *Heterodontus portusjacksoni* in the coastal waters of eastern Australia. *Journal of Fish Biology* 72: 2615–2633.

Venus's Girdle (*Cestum*, Ctenophora)

Matsumoto, G. I. 1991. Swimming movements of ctenophores, and the mechanics of propulsion by ctene rows. *Hydrobiologia* 216–217: 319–325.

Double-Bagged (*Stephalia*, Siphonophora)
Pugh, P. R. 1983. Benthic siphonophores: A review of the family Rhodaliidae (Siphonophora: Physonectae). *Philosophical Transactions of the Royal Society of London B* 301: 165–300.

Catch of the Day (*Bathysaurus*, Osteichthyes)
Sulak, K. J., Wenner, C. A., Sedberry, G. R., and van Guelpen, L. 1985. The life-history and systematics of deep-sea lizard fishes, genus *Bathysaurus* (Synodontidae). *Canadian Journal of Zoology* 63: 623–642.

Pink Paint (*Enypniastes*, Holothuria)
Hartmann, G. and Hertmann-Schröuder, G. 1988. Deep-sea Ostracoda, taxonomy, distribution and morphology. *Developments in Palaeontology and Stratigraphy* 11: 699–707.
Robison, B. H. 1992. Bioluminescence in the benthopelagic holothurian *Enypniastes eximia*. *Journal of the Marine Biological Association of the U.K.* 72: 463–472.

Shape Changers (*Pourtalesia*, Echinoidea)
Mooi, R. and David, B. 1996. Phylogenetic analysis of extreme morphologies: Deep-sea holasteroid echinoids. *Journal of Natural History* 30: 913–953.
Saucede, T., Mironov, A. N., Mooi, R., and David, B. 2000. The morphology, ontogeny and inferred behaviour of the deep-sea echinoid *Calymne relicta* (Holasteroida). *Zoological Journal of the Linnean Society* 155: 630–648.

Coins (*Orbitolites*, Foraminifera)
Beavington-Penney, S. J. and Racey, A. 2004. Ecology of extant nummulitids and other larger benthic foraminifera: Applications in palaeoenvironmental analysis. *Earth Science Reviews* 67: 219–265.

Pearly King (*Nautilus*, Cephalopoda)
Saunders, W. B. and Landman, N. H. Nautilus. 2009. *The Biology and Paleobiology of a Living Fossil*. Springer, Dordrecht.

The Moral of the Mudskipper (*Periophthalmus*, Osteichthyes)
Pace, C. M. and Gibb, A. C. 2014. Sustained periodic terrestrial locomotion in airbreathing fishes. *Journal of Fish Biology* 84: 639–660.

Jaws (*Carcharodon*, Osteichthyes)
Rasalato, E., Maginnity, V., and Brunnschweiler, J. M. 2010. Using local ecological knowledge to identify shark river habitats in Fiji (South Pacific). *Environmental Conservation* 37: 90–97.

Beauty and the Beast (*Basilissa*, Gastropoda); *Typhonus*, Osteichthyes)
Marine Biodiversity Hub (University of Tasmania) Blog post, 31 May 2017.

The Lancelet (*Branchiostoma*, Cephalochordata)
Igawa, T., Nozawa, M., Suzuki, D. G., Reimer, J. D., Morov, A. R., Wang, Y., Henmi, Y. and Yasui, K. 2017. Evolutionary history of the extant amphioxus lineage with shallow-branching diversification. *Scientific Reports* 7: 1–14.

A Deadly Snail (*Conus*, Gastropoda)

Kohn, A. J. 1956. Piscivorous gastropods of the genus *Conus*. *Proceedings of the National Academy of Science of the U.S.A.* 42: 168–171.

Olivera, B. M. 2002. *Conus* venom peptides: Reflections from the biology of species and clades. *Annual Reviews of Ecology and Systematics* 33: 25–47.

Decorator Crabs (Majidae, Decapoda)

Wicksten, M. K. 1992. A review and a model of decorating behavior in spider crabs (Decapoda, Brachyura, Majidae). *Crustaceana* 64: 314–325.

The Inverse Hydra (*Syllis*, Polychaeta)

Crosland, C. 1933. Distribution of the polychaete worm, *Syllis ramosa* McIntosh. *Nature* 131: 242.

Glasby, C. J., Schroeder, P. C. and Aguado, M. T. 2012. Branching out: A remarkable new branching syllid (Annelida) living in a Petrosia sponge (Porifera: Demospongiae). *Zoological Journal of the Linnean Society* 164: 481–497.

Schroeder, P. C., Aguado, M. T., Malpartida, A. and Glasby, C. J. 2017. New observations on reproduction in the branching polychaetes *Ramisyllis multicaudata* and *Syllis ramosa* (Annelida: Syllidae: Syllinae). *Journal of the Marine Biological Association of the U.K.* 97: 1167–1175.

Chapter 13

Spirula (Cephalopoda)

Clarke, M. R. 1970. Growth and development of *Spirula spirula*. *Journal of the Marine Biological Association of the U.K.* 50: 53–64.

Price, G. D., Twitchett, R. J., Smale, C. and Marks, V. 2009. Isotopic analysis of the life history of the enigmatic squid *Spirula spirula*, with implications for studies of fossil cephalopods. *Palaios* 24: 273–279.

Warnke, K. and Keupp, H. 2005. *Spirula*—a window to the embryonic development of ammonoids? Morphological and molecular indications for a palaeontological hypothesis. *Facies* 51: 60–65.

The Garden of Forking Tubes (Komiakacea)

Cartwright, N. G., Gooday, A. J., and Jones, A. R. 1989. The morphology, internal organization, and taxonomic position of *Rhizammina algaeformis* Brady, a large, agglutinated, deep-sea foraminifer. *Journal of Foraminiferal Research* 19: 115–125.

Gooday, A. J., Shires, R., and Jones, A. R. 1997. Large, deep-sea, agglutinated Foraminifera: Two differing kinds of organization and their possible ecological significance. *Journal of Foraminiferal Research* 27: 278–291.

Lecroq, B., Gooday, A. J., Cedhagen, T., Sabbatini, A., and Pawlowski, J. 2009. Molecular analyses reveal high levels of eukaryotic richness associated with enigmatic deep-sea protists (Komokiacea). *Marine Biodiversity* 39: 45–55.

Tendal, O. S., and Robert R. Hessler. 1977. An introduction to the biology and systematics of Komokiacea. *Galathea Report* 14: 165–194.

The Sage of Ternate

Bulmer, M. 2005. The theory of natural selection of Alfred Russel Wallace FRS. *Notes and Records of the Royal Society* 59: 125–136.

Smith, C. H. 2016. Did Wallace's Ternate essay and letter on natural selection come as a reply to Darwin's letter of 22 December 1857? A brief review. *Biological Journal of the Linnean Society* 118: 421–425.

The Cage of Thorns (*Freyella*, Asteroidea)
Emson, R. H. and Young, C. M. 1994. Feeding mechanism of the brisingid starfish *Novodinia antillensis*. *Marine Biology* 118: 433–442.

Eating Wood (*Leptochiton*, Polyplacophora)
Sigwart, J. D. and Sirenko, B. 2012. Deep-sea chitons from sunken wood in the West Pacific (Mollusca: Polyplacophora: Lepidopleurida): Taxonomy, distribution, and seven new species. *Zootaxa* 3195: 1–38.
Sirenko, B. 2004. The ancient origin and persistence of chitons (Mollusca, Polyplacophora) that live and feed on deep submerged plant matter (xylophages). *Bollettino Malacologico* 5: 111–116.
Voight, J. R. 2013. Xylotrophic bivalves: Aspects of their biology and the impacts of humans. *Journal of Molluscan Studies* 81: 175–186.

Chapter 14
The Glass Hotel (*Euplectella*, Hexactinellida)
Saito, T. and Komai, T. 2008. A review of species of the genera *Spongicola* de Haan, 1844 and *Paraspongicola* de Saint Laurent and Cleva 1981 (Crustacea, Decapoda, Stenopodidea, Spongicolidae). *Zoosystema* 30: 87–147.
Saito, T. and Takeda, M. 2003. Phylogeny of the family Spongicolidae (Crustacea: Stenopodidea): Evolutionary trend from shallow-water free-living to deep-water sponge-associated habitats. *Journal of the Marine Biological Association of the U.K.*, 83: 119–131.

A Fierce Scallop (*Propeamussium*, Bivalvia)
Morton, B. and Thurston, M. H. 1989. The functional morphology of *Propeamussium lucidum* (Bivalvia: Pectinacea), a deep-sea predatory scallop. *Journal of Zoology* 218: 471–496.
Temkin, I. and Strong, E. E. 2013. New insights on stomach anatomy of carnivorous bivalves. *Journal of Molluscan Studies* 79: 332–339.

The Challenger Deep
Jamieson, A. J., Fujii, T., Mayor, D. J., Solan, M., and Priede, I. G. 2009. Hadal trenches: The ecology of the deepest places on earth. *Trends in Ecology and Evolution* 25: 190–197.

Huxley's Mistake (*Bathybius*)
Rehbock, P. F. 1975. Huxley, Haeckel and the oceanographers: The case of *Bathybius haeckelii*. *Isis* 66: 504–533.
Rice, A. L. 1983. Thomas Henry Huxley and the strange case of *Bathybius haeckelii*: A possible alternative explanation. *Archives of Natural History* 11: 169–180.
Rupke, N. A. 1976. *Bathybius haeckelii* and the psychology of scientific discovery. *Studies in the History and Philosophy of Science* 7: 53–62.

Chapter 15

Knots in Slime (*Myxine*, Agnatha)

Clark, A. J., and Summers, A. P. 2007. Morphology and kinematics of feeding in hagfish: Possible functional advantages of jaws. *Journal of Experimental Biology* 210: 3897–3909.

Zintzen, V., Roberts, C. D., Anderson, M. J., Stewart, A. L., Struthers, C. D., and Harvey, E. S. 2011. Hagfish predatory behaviour and slime defence mechanism. *Scientific Reports* 1: 131–137.

The Mop-Headed Animal (*Branchiocerianthus*, Hydroida)

Miyajimi, M. 1900. On a specimen of a gigantic hydrid, *Branchiocerianthus imperator* (Allman), found in the Sagaki Sea. *Journal of the College of Science of the Imperial University of Tokyo* 13: 235–262.

Omori, M. and Vervoort, W. 1986. Observations of a living specimen of the giant hydroid *Branchiocerianthus imperator*. *Zoologische Mededelingen* 60: 257–261.

The Bamboo Grove (*Bathygorgia*, Alcyonaria; *Culeolus*, Tunicata)

Lapointe, A. and Watling, L. 2015. Bamboo corals from the abyssal Pacific: *Bathygorgia*. *Proceedings of the Biological Society of Washington* 128: 125–136.

Lescano, M. N., Fuentes, V. L., Sahade, R., and Tatian, M. 2011. Identification of gut contents and microscopical observations of the gut epithelium of the microphagous ascidian *Cibacapsa gulosa* Monniot and Monniot 1983 (Phlebobranchia, Octacnemidae). *Polar Biology* 34: 23–30.

Robison, B. H., Raskoff, K. A., and Sherlock, R. E. 2005. Adaptations for living deep: A new bathypelagic doliolid, from the eastern North Pacific. *Journal of the Marine Biological Association of the U.K.* 85: 595–602.

Diversity at Depth

Kupriyanova, E. K., Vinn, O., Taylor, P. D., Schopf, J. W., Kudryatsev, A. B., and Bailey-Brock, J. 2014. Serpulids living deep: Calcareous tubeworms beyond the abyss. *Deep-Sea Research I* 90: 91–104.

Mironov, A. N., Minin, K. V., and Dilman, A. B. 2015. Abyssal echinoid and asteroid fauna of the North Pacific. *Deep-Sea Research II* 111: 357–375.

Small Fry (Cumacea, Tanaidacea)

Blazewicz-Paszkowycz, M. and Ligowsky, R. 2002. Diatoms as food source indicator for some Antarctic Cumacea and Tanaidacea (Crustacea). *Antarctic Science* 14: 11–15.

Dennell, R. 1934. The feeding mechanism of the cumacean crustacean *Diastylis bradyi*. *Transactions of the Royal Society of Edinburgh* 58: 125–142.

Jones, N. S. and Sanders, H. L. 1972. Distribution of Cumacea in the deep Atlantic. *Deep-Sea Research* 19: 737–745.

Miljutin, D. M., Gad, G., Miljutina, M. M., Moklevsky, V. O., Fonseca-Genevois, V., and Esteves, A. M. 2010. The state of knowledge on deep-sea nematode taxonomy: How many valid species are known down there? *Marine Biodiversity* 40: 143–159.

Wolff, T. 1956. Crustacea Tanaidacea from depths exceeding 6000 meters. *Galathea Reports* 2: 187–241.

An Enigmatic Polyp (*Stephanoscyphus*, Scyphozoa)

Allma, X. 1874. On the structure and systematic position of *Stephanoscyphus mirabilis*, the type of a new order of Hydrozoa. *Transactions of the Linnean Society Series 2 Zoology* 1: 61–66.

Boero, F. 1984. The ecology of marine hydroids and effects of environmental factors: A review. *Marine Ecology* 5: 93–118.

Chapman, D. M. and Werner, B. 1972. Structure of a solitary and a colonial species of *Stephanoscyphus* (Scyphozoa, Coronatae) with observations on periderm repair. *Helgolander wiss. Meeresunters* 23: 395–421.

Jarms, G. 1991. Taxonomic characters from the polyp tubes of coronate medusa (Scyphozoa, Coronatae). *Hydrobiologia* 217: 463–470.

Chapter 16

The Admiralty Worm (*Synapta*, Holothuria)

Woodward, S. P. and Barrett, L. 1858. On the genus *Synapta*. *Proceedings of the Zoological Society of London* 1858: 360–367.

The Beauty of the Deep (*Bathydoris*, Nudibranchia)

Wagele, H. 1989. A revision of the Antarctic species of *Bathydoris* Bergh, 1884 and comparison with other known bathydorids (Opisthobranchia, Nudibranchia). *Journal of Molluscan Studies* 55: 343–364.

More Strange Lumps (*Stannophyllum*, Xenophyophorea)

Riehl, T., Bober, S., Voltski, I., Malyutina, M. V., and Brandt, A. 2017. Caught in the act: An abyssal isopod collected while feeding on Komokiaceae. *Marine Biodiversity* 10: 1007–1008.

Glass Mines (Radiolaria)

Gowing, M. M. 1989. Abundance and feeding ecology of Antarctic phaeodarian radiolarians. *Marine Biology* 103: 107–118.

Matsuoka, A. 2007. Living radiolarian feeding mechanisms: New light on past marine ecosystems. *Swiss Journal of Geoscience* 100: 273–279.

Nakamura, Y. and Suzuki, N. 2015. Phaeodaria: Diverse marine cercozoans of world-wide distribution. Pp. 223–249 in Ohtsuka, S., Suzaki, T., Horiguchi, T., Suzuki, N., and Not, F. (editors), *Marine Protists: Diversity and dynamics*. Springer, Japan: Tokyo.

The Death of a Naturalist

Kortum, G. 1996. The German challenge of Neptune: On the short life and tragic death of Rudolph von Willemoes-Suhm (1847–1875) and a memorial plate on graveyard in Holstein. *Newsletter of the International Union of the History and Philosophy of Science* 8: 3–6.

Wyville Thomson, C. 1875. Dr R. von Willemoes-Suhm. *Nature* 13: 88–89.

Benthesicymus (Decapoda, Natantia)

Jamieso, A. J., Fujii, T., Solan, M., Matsumoto, A. K., Bagley, P. M., and Priede, I. G. 2009. First findings of decapod crustacea in the hadal zone. *Deep-Sea Research I* 56: 641–647.

Fire Coral (*Millepora*, Hydrozoa)

Lewis, J. B. 2006. Biology and ecology of the hydrocoral *Millepora* on coral reefs. *Advances in Marine Biology* 50: 2–55.

Bathysaurus (*Oneirophanta*, Holothuria)

Ramirez-Llodra, E., Reid, W. D. K., and Billett, D. S. M. 2005. Long-term changes in re-productive patterns of the holothurian *Oneirophanta mutabilis* from the Porcupine Abyssal Plain. *Marine Biology* 146: 683–693.

A Homeless Crab

Lemaitre, R. 1998. Revisiting *Tylaspis anomala* Henderson, 1885 (Parapaguridae), with comments on its relationships and evolution. *Zoosystema* 20: 289–305.

Whalefall

Dahl, E. 1979. Deep-sea carrion feeding amphipods: Evolutionary patterns in niche adap-tation. *Oikos* 33: 167–175.

Jamieson, A. J., Fujii, T., Sola, M., Matsumotu, A. T., Bagley, P. M., and Priede, I. G. 2009. Liparid and macrourid fishes of the hadal zone: In situ observations of activity and feeding behaviour. *Proceedings of the Royal Society B* 276: 1037–1045.

Nebaliopsis (Crustacea, Leptostraca)

Hessler, R. R. and Schram, F. R. 1984. Leptostraca as living fossils. Pp. 187–191 in Eldredge, N. et al., (editors), *Living Fossils*. Springer-Verlag, New York.

That Sinking Feeling

Alldredge, A. L. and Silver, M. W. 1988. Characteristics, dynamics and significance of ma-rine snow. *Progress in Oceanography* 20: 41–82.

Bochdansky, A. B., Clouse, M. A., and Herndl, G. J. 2017. Eukaryotic microbes, princi-pally fungi and labyrinthulomycetes, dominate biomass on bathypelagic marine snow. *ISME Journal* 11: 362–373.

Plume and Splash

Higgs, N. D., Gates, A. R., and Jones, D. O. B. 2014. Fish food in the deep sea: Revisiting the role of large food-falls. *PLoS One* 9: e96016.

Smith, C. R. and Baco, A. R. 2003. Ecology of whale falls at the deep-sea floor. *Oceanography and Marine Biology* 41: 311–354.

Smith, C. R., Glover, A. G., Treude, T., Higgs, N. D., and Amon, D. J. 2015. Whale-fall ecosystems: Recent insights into ecology, paleoecology and evolution. *Annual Review of Marine Science* 7: 571–596.

Extraterrestrials

Dekov, V. M., Molin, G. M., Dimova, M., Griggio, C., Rajta, I., and Uzonyi, I. 2007. Cosmic spherules from metalliferous sediments: A long journey to the seafloor. *Neues Jahrbuch fur Minerologie—Abhandlung* 183: 269–282.

Smith, C. R. and Demopoulos, A. W., 2003. The deep Pacific Ocean floor. Pp. 179–218 in Tyler, P.A. (editor), *Ecosystems of the Deep Oceans*. Elsevier Science, New York.

Uskinowicz, G. 2012. Spherical, magnetic grains of extraterrestrial origin, isolated from Pacific sediments. *Oceanological and Hydrobiological Studies* 41: 48–53.

The Waters AboveBurghart, S. E., Hopkins, T. L., and Torres, J. J. 2007. The bathypelagic Decapoda, Lophogastrida and Mysida of the eastern Gulf of Mexico. *Marine Biology* 152: 315–327.

Mauchline, J. 1972. The biology of bathypelagic organisms, especially Crustacea. *Deep-Sea Research* 19: 753–780.

The Usual Suspects
Frankenberg, D. and Menzies, R. J. 1968. Some quantitative analyses of deep-sea benthos off Chile. *Deep-Sea Research* 15: 623–626.

Chapter 17
Animal Forests (Gorgonians)
Buhl-Mortensen, P., Buhl-Mortensen, L., and Purser, A. 2016. Trophic ecology and habitat provision in cold-water coral ecosystems. Pp. 1–26 in Rossi, S. (editor), *Marine Animal Forests*. Springer, Cham, Switzerland.

A Very Unexpected Result (*Cephalodiscus*, Pterobranchia)
Dilly, P. N. 1993. *Cephalodiscus graptiloides* sp. Nov. a probable extant graptolite. *Journal of Zoology* 229: 69–78.
Lester, S. M. 1985. *Cephalodiscus* sp. (Hemichordata: Pterobranchia): Observations of functional morphology, behavior and occurrence in shallow water around Bermuda. *Marine Biology* 85: 263–268.
Mitchell, C. E., Melchin, J. A., Cameron, C. B., and Maletz, J. 2012. Phylogenetic analysis reveals that *Rhabdopleura* is an extant graptolite. *Lethaia* 46: 34–46.
Ridewood, W. C. 1921. On specimens of *Cephalodiscus densus* dredged by the "Challenger" in 1874 at Kerguelen Island. *Annals and Magazine of Natural History* 46: 433–440.

The Angler (*Atolla*, Scyphozoa)
Helm, R. R. 2018. Evolution and development of scyphozoan jellyfish. *Biological Reviews* 93: 1228–1250.
Hunt, J. C. and Lindsay, D. J. 1998. Observations on the behavior of *Atolla* (Scyphozoa: Coronatae) and Nanomia (Hydrozoa: Physonectae): Use of the hypertrophied tentacle in prey capture. *Plankton Biology and Ecology* 45: 239–242.

The Espalier Animals (Hydrozoan stony corals: Stylasteridae)
Cairns, S. D. 2011. Global diversity of the Stylasteridae. *PLoS One* 6: e21670.

Chapter 18
Renilla (Anthozoa, Cnidaria)
Kastendiek, J. 1976. Behavior of the sea pansy *Renilla kollikeri* (Coelenterata: Pennatulacea) and its influence on the distribution and biological interactions of the species. *Biological Bulletin* 151: 518–537.
Yuan, M. et al. 2017. Prolonged bioluminescence imaging in living cells and mice using novel pro-substrates for *Renilla* luciferase. *Organic and Biomolecular Chemistry* 15: 10238.

The Vampire Snail (*Pisolamia*, Gastropoda)
Bouchet, P. 1976. *Pisolamia*, nouveau genre de Gasteropode, parasite de l'Holothurie abyssale *Oneirophanta mutabilis*. *Comptes rendues des seances de l'Academie des sciences D* 282: 1013–1016.
Bouchet, P. and Lutzen, J. 1980. Deux gastéropodes parasites d'une holothurie Élasipode. *Bulletin du Muséum Nationale d'Histoire Naturelle de Paris* 4ᵉ, series 2: 59–75.

Salazar-Vallejo, S. I. and Zhadan, A. E. 2007. Revision of *Buskiella* McIntosh, 1885 (including Flota Hartman, 1967) and description of its trifid organ (Polychaeta: Flotidae). *Invertebrate Zoology* 41: 65–82.

The Free Rider
Mosher, C. V. and Watling, L. 2009. Partners for life: A brittle star and its octocoral host. *Marine Ecology Progress Series* 397: 81–88.

The Last Clam (*Vesicomya*, Bivalvia)
Krylova, E. M. and Sahling, H. 2010. Vesicomyidae (Bivalvia): Current taxonomy and distribution. *PLoS One* 5: e9957.
Krylova, E. M., Sahling, H., and Borowski, C. 2018. Resolving the status of the families Vesicomyidae and Kelliellidae (Bivalvia: Venerida) with notes on their ecology. *Journal of Molluscan Studies* 84: 69–91.
Von Cosel, R., Salas, C., and Hoisaeter, T. 2001. Vesicomyidae (Mollusca: Bivalvia) of the genera *Vesicomya*, *Waisiuconcha*, *Isorropodon* and *Callogonia* in the eastern Atlantic and the Mediterranean. *Sarsia* 86: 333–366.

Chapter 19
Figurehead of HMS *Challenger*.
Image on website of *Royal Museums Greenwich*, FigID F0287.

Chapter 20
Anon. 1885. Obituary. Dr W.B. Carpenter. *The Lancet* 2: 928–929.
Bourne, G. C. 1892. Memoir of Henry Nottidge Moseley. In second edition of Moseley, *Notes by a Naturalist* op.cit.
Burstyn, H. L. 1975. Science pays off: Sir John Murray and the Christmas Island phosphate industry, 1886–1914. *Social Studies of Science* 5: 5–34.
Deacon, M. and Savours, A. 1976. Sir George Strong Nares (1831–1915). *Polar Record* 18: 127–141.
Leyton, L. 1969. Sir Charles Wyville Thomson (1820–1882) Sir James Murray (1841–1914) The Challenger expedition. Pp. 1–30 in Harré, R. (editor), *Some Nineteenth Century British Scientists*. Pergamon, London.
Manten, A. A. 1972. C. Wyville Thomson, J. Murray, and the *Challenger* expedition. *Earth Science Reviews* 8: 255–266.
Mill, H. R. 1925. Obituary. Mr J.Y. Buchanan, FRS. *Nature* 116: 719–720.
Redfern, P. 1888. Sir Charles Wyville Thomson FRSS L. and E. *Proceedings of the Royal Society of Edinburgh* 14: 58–80.
Sexton, R., Sexton, C., and Mackay, K. 2008. Sir John Murray of the *Challenger* expedition: Founder of oceanography. *Forth Naturalist and Historian* 31: 15–34.
Stein, G. M. 2006. The Challenger Medal roll (1895). *Journal of the Orders and Medals Research Society* 45: 241–248.
Wharton, W. J. L. 1888. Account of Christmas Island, Indian Ocean. *Proceedings of the Royal Geographical Society* 10: 613–624.

Chapter 21

Burkhardt, F. and Secord, J. A. (editors). 2016. *The Correspondence of Charles Darwin.* Appendix V. "Letters regarding the HMS Challenger specimens." Cambridge University Press, Cambridge.

Lingwood, P. F. 1981. The dispersal of the collections of HMS *Challenger*: An example of the importance of historical research in tracing a systematically important collection. *Archives of Natural History* 1: 71–77.

Low, M. E. Y. and Evenhuis, N. L. 2013. Dates of publication of the *Zoology* parts of the *Report of the Scientific Results of the Voyage of HMS Challenger during the years 1873–76. Zootaxa* 3701: 401–430.

Roemmich, D., Gould, W. J., and Gilson, J. 2012. 135 years of global warming between the *Challenger* expedition and the Argo programme. *Nature Climate Change* 2: 425–428.

General Index

Note: oceans, continents and most large countries are not indexed, nor is the voyage itself.

Index of Animals

Passing references to common names are not indexed.